I0070710

ISBN: 978-0-578-38452-8 (Paperback)
Revised on: 10/30/22

Library of Congress Control Number: 2022904255

The events in this book are the author's memories and from his perspective. Certain names have been changed to protect the identities of those involved.

All the information provided in this book is for information purposes only and is not to be taken as legal or financial advice.

Printed by Lulu Press Inc; Research Triangle, NC 27709 United States of America. First printing edition 06-2022.

Smarter Investments Corp
Fabian Videla
9218 Cypress Green Dr.
Jacksonville, FL 32256
fabianvidela.com

DEDICATIONS

There have been many people who've influenced me in positive and uplifting ways throughout my life. Without a doubt, one of the most influential was my Grandma Julia. Without her teachings and sacrifices in my early and formative years, I'd be a very different person today.

This book also pays homage to my Aunt Gladys, who chose to build her life with my siblings and me at the center of it. We'll never forget the tremendous sacrifices she made for us.

I'd also like to honor my Mom, who showed me what unconditional love is and who never misses an opportunity to demonstrate her love for me. I can't express in words how much I love her!

And there are so many more people to whom I'd like to dedicate this book:

To my "querido viejo"—a nod to my Dad through a song by Piero that he loved. I miss you lots, Dad!

To Antonio, who was instrumental in my early adult years to put me in the right path and supported me and my family without ever asking for anything in return.

To Lucy, my mother-in-law, who relentlessly and tirelessly fought to protect her family and pushed me to become a better man, a better husband, and a better father.

To my beloved uncles Francisco, Juan, Ricardo, Eusebio, and Carlos, who are no longer with us but live on in my heart.

To my uncles Alfredo, Daniel, Jorge and Guillermo. To my aunts Ramona, Coca, Chiquita, Silvia, Mabel, Rita, Mary, Rosi and Estela who among many others taught me many lessons and gave me much loving guidance and support.

To my brother Javier, who is my idol in so many ways and who enables me to help many people at a distance. He perfectly embodies the kind of man who will instantly make you feel like an old friend, if not family. Even strangers feel pride in knowing him.

To my friend Daniel, who gave me my first job in America and, together with his wife Rosa, opened his home when my family and I had no place to live.

To my children, Alexa, Karen, Andrew, and Kiara, who taught me how to be a father, challenged me to be a better man, and always inspired me with their unconditional love. Your existence has changed my life for the better; every day my life improves because you are all in it.

To my friend Gustavo, who's been my unconditional best friend since we were kids. He demonstrates time and again what loyalty is.

To my friends Steve and Maria, who opened not only their home but also their hearts to my family and me on a cold Thanksgiving Day many years ago. They adopted us as their own family and never let us forget how much we are loved and mean to them.

To the team at Smarter Remodeling, whose talents and dedication have allowed me to step away from the day-to-day work to pursue my other goals. Special thanks to Bill who pushed me to not take the easy path in completing this book.

To Paul, who believed in what we could accomplish together and trusted my vision from the very beginning.

To Ed, who has been a reliable sounding board for my ideas and provided me with sage counsel and guidance.

Finally and most importantly, to the person who's never doubted that I'd become successful. The person who's helped shape and transform our company into what it is today. To the person who has loved me unconditionally despite all my faults. To the one who embraces my shortcomings and makes them strengths. To my beloved, who taught me to be a husband and a father. Without you, this book would have never happened. I dedicate this book to my wife, Agata.

CONTENTS

INTRODUCTION

"Today I will do what others won't, so tomorrow I can do what others can't."

~Jerry Rice

I believe my willingness to live by that phrase—even as a child (when Jerry Rice was still in college)—has been critical to my success.

Since a young age, I've been relentless in my pursuit of success and wealth. My definitions of both have changed greatly over time, but I have never stopped pursuing them.

I don't define success and wealth in terms of only money anymore. As I've matured, I've learned there are things much more important than money. Still, I can't let go of my hyper-competitive spirit, so I admittedly still use it to keep score.

My story is one that's been told many times already. It's a story of rags to riches. Of triumphs and failures. Of courage and fear. Of laughter and tears. It could be your story, too, if your desire is strong enough.

My goal in writing this book is to tell my story in a way that shows how the events of my life and the actions I took taught me skills and lessons that I later applied to build my success. I learned from every triumph and every failure. Contrary to what many people around me think, I wasn't fearless when taking the chances I did. I was afraid. But I had the courage to do it anyway because courage is not the absence of fear but rather the presence of mind to act even when you are afraid.

As Franklin D. Roosevelt said, "Courage is not the absence of fear but rather the assessment that something else is more important than fear." To me, that something *more important* was to succeed. Not just for myself, but for my family, my friends, and my community.

Taking action, even when I was afraid, has been one of the most important things I have learned over the course of my life. Since childhood, I haven't been able to cope with being afraid; so when something scares me, I have to face it at least once so I can move on. I never let fear control my actions. Rather, I let reason control them—and I encourage you to do the same. Don't let fear stop you from seeking greatness and achieving it.

The fear of failure is the most common cause why the vast majority of potential entrepreneurs never take action. Often, they believe failing will make them a failure. In reality, failing teaches how to succeed. You wouldn't be reading this book if Thomas Edison had stopped the first time he failed to create a light bulb or if Alexander Bell decided to call it quits when he failed to create the telephone on the first try.

Ultimately, I hope this book inspires others to take action—to do something, no matter how small, to improve themselves or their business. And then to do it again. And again. And again. This is the formula of how I went from being a nameless handyman to being the owner of a multi-million-dollar construction company that's been part of the '50 Fastest Growing Companies' in North Florida three times since 2017 in addition to being included in the Inc 5000, for 2022, which lists the fastest growing companies in the country.

These days, I dedicate most of my time to my new venture: **Smarter Remodeling 360 Solutions**, a company that will revolutionize the way contractors manage and understand their business through the use of proprietary software and industry-specific training that will empower them to create more efficient and profitable enterprises.

I created this new company while my construction business kept running along without me, all thanks to the systems and procedures we created over the last two decades and the lessons I learned along the way.

I challenge you to find one lesson or one story in this book that you can use to better yourself or your company or to help somebody else. A minimum of one. If you do, drop me a line and tell me how you used the lesson to positively impact something or somebody. It will make me extremely happy to know I was able to touch someone's life, even if only in a small way.

In all honesty, that has always been my dream: to meaningfully affect other people's lives. It's my desire that the lessons others have taught me transcend their own plane of existence, living on and touching the lives of other people for generations to come. I deeply believe this is the best way to honor the people in our lives who inspired us and who made a difference: by passing those teachings on to others.

Each chapter in this book highlights key lessons that I have learned about that chapter's topic. The *Lessons of Success*, summarizes a helpful takeaway from a story shared in the chapter. My hope is that you'll be able to come back to the book and quickly go through the *Lessons of Success* to refresh your memory and put you back on track to producing the changes you need in life and business. These lessons can be found in the final chapter: Lessons of Success.

Finally, you can access additional content by scanning the QR Codes that you find throughout the book. They will take you to my website where you will need to register for a free account and then you can access all the additional content.

So go ahead. Learn. Thrive. Teach. Live. **Succeed**.

SCAN ME FOR A VIDEO MESSAGE

PROLOGUE

I walked into the Porsche dealership with my son Andrew by my side, as always. He is my unconditional buddy, and together we've traveled the world and embarked on many adventures.

My wife Agata and I were blessed with our son Andrew the year we arrived in the USA. He is a special kid (he's not a *kid* anymore, but he'll always be a kid to me), and *special* has never been used more appropriately: he truly is exceptional.

Andrew loves and cares for people in a way that's bewitching. He laughs at everything and doesn't worry about what others think of him. He loves sports and is such an expert that we can ask him about any score we might want to know about—no Googling necessary! Even today, Andrew believes in Santa Claus, and he'll defend the red-suited man with a passion! But that doesn't surprise me, as I still believe in Santa Claus, too, although perhaps not in the sense kids do. I believe in the spirit that descends upon us during the Christmas season—the one that inspires us to be kinder to each other and more compassionate with the less fortunate.

In many ways, Andrew and I are very similar. For example, like me, he's not too fond of the word "no."

Well, there is an exception. When you ask him to take a shower, "No" is certainly a prominent part of his vocabulary. But that may be genetically imprinted, as I never liked showering when I was his age!

Gazing at the cars, Andrew pointed to the red Porsche Panamera and said, "That's my car, Daddy."

Smiling, I looked at him and replied, "Slow down, buddy. You don't even have a driver's permit yet, but maybe one day. For now, let's find Daddy a car."

We walked around the lot and a salesman joined us, touting the brand's virtues. He was a little arrogant but Andrew was listening to him very closely. I was thinking perhaps customers driving pickup trucks didn't normally buy these cars; I guess he didn't realize my truck was a top-of-the-line luxury Ford pickup truck that cost more than many of the cars on the lot.

I asked the salesman for the price of the red Panamera and he responded with the monthly payment. I looked at him and said, "I'm not interested in payments, just the cash price." He looked annoyed and replied, "This particular one is right under $100,000." I nodded and walked around. He didn't offer to show the car or to go for a drive. So, I said to Andrew, "Let's go, bud. I don't like it here." Andrew, being the kind person he is, turned around, smiled at the salesperson and said goodbye. I left without saying a word.

The salesperson looked surprised and was probably relieved he wouldn't have to waste his time. I wasn't interested in a Porsche anyway; for a brief moment I considered coming back with a new car, slowly driving in his parking lot and saying to him, "You work on commissions, right? Big mistake, big mistake"—just like in the movie *Pretty Woman*. But I'm not pretty or a woman, so I chose to walk out of there quietly. Andrew wouldn't have allowed me to say that anyway; he is too nice to hold grudges. On the other hand, me … not so much.

My dream car has always been a BMW 7 Series with the executive package and the M3 sport configuration. I had one in mind at the BMW dealership just down the block, so Andrew and I headed that way. Immediately upon our arrival, a young salesman greeted me, complimented my truck, and offered Andrew and me some cold water, as the day was hot. We accepted, and then I proceeded to mention what I was looking for.

The salesman was quick to pay a compliment. "A refined taste in cars. Love it," he said. "Come on over here and we'll look at the Series 7 cars we have available."

We moved to another section of the lot and there they were. It was like I stepped into heaven. They looked amazing all lined up in a beautiful row; I swear I could see the front grills of those cars talking to me like little puppies saying, "Take me! Take me!"

I walked up to a white Series 7. The young salesman opened the doors and then excused himself to retrieve the keys. I stepped inside and was blown away by its luxury. The level of detail was incredible. The black leather had a diamond pattern and felt so supple and lush. I was in love. The salesman came back and offered me the keys.

This time, he was with an older gentleman who introduced himself as the General Manager who thanked me for taking the time to look at his cars. He offered to take us for a test drive and I told him, "No need, I want it." He was surprised but said, "Fantastic! Would you like to drive it so we can show you all the features? Or perhaps you'd like to look at some of the other Series 7s and see if there's another one you'd like better?"

I thanked him and said I had been waiting to be able to afford this car for many years, and now that I could have it, I didn't want to waste any time testing it. Without a doubt in my mind, I knew I wanted it. "Give me your best price," I said. "Let's close the deal now."

He looked at the sticker, got on his phone, and ran some numbers. He then said, "Sticker price is $141,000, but so we can avoid haggling over price, I'll sell it to you for $123,000."

I looked at him and asked, "Would you take my American Express or do you prefer a check?" He shook my hand and replied, "Whatever you prefer, Mr. Videla."

I closed my eyes and the movie that had played in my mind since I was a child started rolling once more. There I was, seated in a luxury car—with a gorgeous blond swimsuit model next to me—wearing designer clothes and a Rolex watch (one of the many in my collection), letting the Ultimate Driving Machine roar along the highway toward my large house with a pool with the feeling that I had made it.

Over time, my dream came true. My wife wasn't a blond swimsuit model (though she could be if she wants to be!) but she is still drop-dead gorgeous and the best gift that has ever been given to me.

Like in the movies, the nerd got the beautiful girl who was way out of his league. I became a luxury-watch collector and drove several high-end cars, including the likes of McLaren, Ferrari, and Lamborghini, although my favorite still is my BMW Series 7.

I became a successful entrepreneur and built a company that made positive changes in the lives of its stakeholders. But, most importantly, my wife and I built a family that has made me the richest man on the planet—and I wouldn't change that for anything in the world.

Before I could achieve the level of success that would allow me to carry a metal card with enough buying power to purchase a luxury car on the spot or the capacity to write a check for its value, I had to learn many lessons and lose everything I had several times.

I had to start over more times than I cared to admit. Interestingly, this puts me in fine company. Any billionaire worth his/her salt has gone bankrupt at least once.

My failures have made me resilient, and they've taught me that nothing in life is guaranteed. Nothing. Well, only death. And that's why, as Muhammad Ali is often quoted, we ought to "live each day like it's your last … one day you're going to be right."

I left my life and extended family in Argentina to come to the USA. I had to work very hard but also very intelligently. Ultimately, my sacrifices would pay off, enabling me to create a company from scratch that would become a model to follow for other construction entrepreneurs around the country.

If you would like to know how I did it, proceed to read. If not, thanks for purchasing this book as I still need to put gas in my Bimmer; the gas prices have been insane lately!

My most sincere hope is that I can inspire your inner entrepreneur to get out and make the changes you need to make to become *your* success story. There is nothing out there but yourself stopping you from becoming a successful entrepreneur. So, get out of your own way and go make some splashes. I'm rooting for you.

Lesson of Success

— 1 —

Customer experiences and kindness.

Be kind like my son Andrew, who did this with the not-so-nice salesman from the Porsche dealership.

Strive to provide exceptional customer experiences, just as the salesmen from the BMW dealership. They went above and beyond to make me feel appreciated during and after the purchase. Therefore, I have remained a loyal customer to them over the years and I speak about them positively all the time. I also frequently speak about the other salesman—but not in a positive light.

Go beyond customer service and embrace creating customer experiences.

CHAPTER 1

---◆---

THE AMERICAN DREAM

For myself and the millions of immigrants before me, coming to America was the opportunity to share in the *American Dream*. Those two words have been part of daily life for Americans for centuries. It's the basic principle that with hard work, anyone can achieve anything, have everything, and be whoever they want to be. For the vast majority of the world, the American Dream seems like a utopia. It's something we often see in movies but in reality it's not attainable. After all, how can it be possible that just by working hard you can have anything you want?

Millions of immigrants arrive in the USA and a short time later believe they're living the American Dream; a few years after arriving they truly believe they have achieved it. And why wouldn't they believe this? They come from countries with rampant poverty and basic life necessities, such as eating every day, are a challenge for many. Having a steady and decent-paying job is an accomplishment of monumental proportions; their annual income is around $15,000 (and in some Latin American countries this is considered a high paying job). Having a car (no matter its age) is a luxury and having a newer model is a symbol of status. Owning your house is the pinnacle of success for the average person.

Compare this to the average American. The average salaried American makes $50,000/year. Eating every day is a fact of life and eating out is a common occurrence. Having a car is not a luxury but a necessity and most families have more than one car (and they are relatively new models). And as for home ownership, over 65% of Americans own their house.

From these figures it's easy to see why for many immigrants coming to America and making $25,000 a year is a dream come true. If they can make it into the "mainstream" and make $40,000 to $50,000, buy a decent car and a house, then it feels like they have achieved the American Dream.

But have they? Is this all it is? It may be enough, but can we get more?

The answer is *Yes*. We can. I did.

Lesson of Success

— 2 —

Live your own American Dream.

No matter what you think the American Dream is, make plans and take actions so you get to live your own version of it.

The beauty of living in the Land of the Free is we get the opportunities to make our dreams a reality. Don't let them pass. Summon the courage, take a leap of faith, and build your American Dream.

CHAPTER 2

THE BEGINNING

Like all stories, mine has a beginning. It starts in Mendoza, Argentina where I was born to a working class, tight-knit family. I'm not sure how my family lived when I was born.

Some of my earliest memories are of when I was about six years old and we moved to a new house. It was huge! We moved from a house built with mud bricks (known as adobe) to a house built with real bricks! The house had concrete slabs, indoor plumbing, and one and a half bathrooms with a water heater (which meant no more bathing with water heated in a pot). It was like I was one of the characters from the Beverly Hillbillies.

There were four bedrooms in the new house, but it was home for twelve people. At first there were just a handful of people in the house, but some of my siblings arrived later and then after that a few cousins as well. It was crowded for a while and I loved it. My entire family lived there and on the weekends we would have huge family dinners.

All the adults worked outside of the home with the exception of Mom who was a homemaker.

Like my son Andrew, when I was young I'd come up with elaborate schemes to avoid showering. I remember my battles with Mom as she pleaded for me to take a shower. Even when I was covered in dirt from playing in the field across the street that we used as our improvised soccer pitch, I insisted that a shower was unnecessary.

One day, I splashed some water onto my face and arms and I was done. But Mom would have none of it, throwing her arms up in the air in complete disbelief. "Fabian, get into the shower right now or I'll get my chancla."

The chancla was not to be trifled with. It was a powerful device that inflicted just the right amount of pain to be feared; and in the hands of a trained operative like Mom, it felt lethal. It seems as though every woman in Latin America gets sent to the Chancla Seal Academy. In all reality, the chancla is nothing but a rubber flip-flop, but ask any kid who grew up in the Hispanic culture and they will attest to the power of the chancla to subdue one into surrendering.

My rear end shivered at the mention of the chancla. I capitulated and ran to the bathroom. Once inside and after turning on the shower, the fear of the chancla subsided and my laziness took over. I sat on the toilet seat and read comics.

But Mom wasn't easily fooled. She could tell I wasn't showering because the water immediately hit the floor. "Fabian!" she yelled at me. "I'm breaking the door down and will leave your butt looking like a baboon if you don't get in the shower right now!"

The fury in Mom's voice was disarming and I gave in. I stretched my hand out and placed it under the running water to check the temperature.

But then, a little devil suddenly appeared on my shoulder and whispered something into my ear, "Did you realize the water is not hitting the floor when only your hand is in there?" I smiled, knowing where this was going. I left my hand there until the water started running cold, and then I turned it off. I splashed some water onto my hair, wrapped myself in a towel, and walked out of the bathroom.

Mom was walking in the hallway and looked at me. I greeted her and said, "I'm done. Clean like a baby." I pulled the towel off my shoulders and showed her. She was about 15 feet away from me but I could see her face transform from that of a beloved sweet mom to that of a ninja mom. In a swift, fluid motion the chancla made its way from her foot to her hand and without missing a beat she launched it towards me, hitting me perfectly in the back of my head as I was turning. I knew I couldn't escape the chancla, as her throw was as precise as a guided laser missile. But I tried anyway. I twisted my head just enough to receive the blow on the back of my head instead of between my eyes where it was intended.

Ninja Mom didn't move an inch and with a voice that instilled fear, she said, "Bring your butt here right now." She paused, then added, "And bring me my chancla."

I knew better than to avoid bringing the chancla back. Based on previous interactions with that chancla, I knew that either I could return it to its legitimate owner or I would suffer the consequences. I picked up the chancla and slowly started walking towards Mom.

"What did I do now?" I asked, holding back tears. I knew the chancla was never deployed without a reason, but surely in this time I did nothing that merited its use! Or so I thought. When I got closer to Mom, she took the towel, pointed to my body and said, "What were you doing in the bathroom? I told you to take a shower."

I was confused. How could she possibly know that I didn't really shower? "I did, Mom," I replied. She asked me to turn around and I feared the chancla was coming back into action. Luckily, she didn't use it to hit me; she used it to point to areas on my back, from shoulders to ankles, saying, "Well, I guess our water is not as wet as it used to be because you have streaks of dirt all over you." Then I realized I couldn't get out of this—I was busted.

When the water had splashed out of my hand, it had left streaks of dirt over my body. I was speechless. I had a few words prepared for that devil on my shoulder, who had now vanished. He wasn't a dummy. Nothing escapes the chancla—as my butt was about to learn. I got a couple of hits within milliseconds of each other as Mom commanded me to go take a shower and leave the door open so she could see I was actually showering. This time I had no choice but to shower with cold water as I had previously wasted the hot water supply.

I learned my lesson, but not without some input from the little devil on my shoulder that mysteriously reappeared: "Next time, watch for the splashes." Fortunately, I knew better than to listen to him again. This time, an angel appeared, reminding me that nothing was worse than to be at the receiving end of a chancla's wrath and a mother's disappointment.

*****

Our house had a patio and it was there that I dreamed my childhood fantasies: my dreams and fantasies were very detailed. I was an expert martial arts fighter who rescued ladies in distress. I built my spaceships to explore other worlds. I was a world-famous soccer player. When I was a soccer player, I pictured myself wearing full gear: a shiny new jersey with the number nine and my name on it, black pants with Adidas soccer cleats and pristine white socks.

I'd be in the stadiums, with the crowds cheering my name. I'd mimic the actions of players in pain, throwing myself on the ground and grabbing my legs like I was in pain, later limping and continuing to play "through the pain" to become the instant hero that took my team to the championship, even when injured. No matter what I was dreaming, I was very good at something that would take me places later in life.

I had a very active imagination which would later be fertilized with the ideas and words of great authors to become a productive ground for my own success. I learned that if I wanted something and I worked hard enough I could get it. It was good to dream, but my dreams had to be very detailed, so precise that I could make them become realities by reaching one milestone at a time.

I watched movies when I was a kid which gave birth to my American Dream. I wanted my own luxury cars and a house with a pool and garden. I always loved watches and I wanted my own luxury watches. I wanted a beautiful wife, too. No one imagined a child from my neighborhood could achieve all those things, but I'd grow up to defy their expectations, becoming the CEO of a multi-million dollar company with all those things, including a beautiful wife and family.

The way I have achieved everything in my life was by first dreaming it. Many people have told me, "Fabian, stop dreaming and get real. That kind of life is not for you. Just get a good paying, stable job."

Regardless of country and circumstance, this was the line of thought many parents told their children. Parents often want better for their children than they had, while hoping they will avoid some of the huge potholes of life. But I kept seeing what *I* wanted. I continued with my detailed dreams—I could visualize every color, every smell, every taste.

Don't stop dreaming. Make your dreams the blueprints for what you want your life to be. Visualize your dreams and then take action. Visualizing without action is just dreaming. Action without visualizing is wasted movement. They must be done together.

Taking action is crucial. Nothing will happen if you just wait for things to come to you. It doesn't matter how big or small your dream is; the most important thing you can do is to take the first step. And then a second step. Keep taking steps towards your goal. Don't stop. Don't rest. Just keep pushing forward.

When you have a setback, learn from the experience and get ready to take the next step to put your goals and dreams back on track.

Always remember, only you can decide when it's enough. Only you can decide how big your dream is. To be on the way to achieving your dreams you only need to do one thing: take the first step.

Lesson of Success

— 3 —

Be careful who you lend your ear to.

In the story above, the "little devil" gave bad advice, and since it was the easiest thing to do, I followed it.

Oftentimes, the path of less resistance is not the right one to follow.

Surround yourself with people you can learn from and whose sage advice you can follow. Beware of false "gurus" or "little devils," and remember, there are no shortcuts for doing the right thing.

Lesson of Success

— 4 —

Dream big.

I encourage you to dream. But dream with an abundance of details. Break the details of your dreams down to a list that can be used to track your progress towards achieving your dream.

But most importantly remember to take action. One small step and then another. Dreams with actions become reality.

CHAPTER 3

MY FAMILY

My grandma was an inspiration to everyone around her and has been one of the most influential people in my life. She was honest to the core and very proud as well.

Grandma never finished elementary school and was a housekeeper for a few "rich" families: they lived in houses with two or more full bathrooms, had real floors, and— most importantly— had cars! She worked long hours performing backbreaking housekeeping tasks. There was no stigma against her working as a housekeeper; it was the way she fed her family.

Grandma made her money stretch by walking miles to save a few bucks buying groceries at different places. She knew a good deal when she saw it and would always bargain, pushing the price down even further. Occasionally she would complain about how expensive the groceries were or how she had walked to three different grocery stores until she found the deal she wanted.

Her labors as a housekeeper and excess amounts of walking made her back and legs hurt and often quite tired. But no matter how she felt, she would wake up each morning and do it all over again.

Sometimes I would go with her to carry the groceries home. On these occasions I got to see how the "other half" lived. Grandma wouldn't miss an opportunity to remind me how important it was to study diligently so that in the future I could be one of the "other half": she didn't want me cleaning houses like she did. She said cleaning was honest work but I could do better than she did.

She was short tempered and my siblings and I would annoy her, so she would react with one of her famous phrases. Our favorite was when she would get angry and attempt to call us by our names but would name everybody in the family before she would get to the name she wanted and then finish it with an expletive such as: "You stop it! Now! I'll get my chancla! Stop! Jani, no, you! Quique! No, you, Guille! Rami! You *shithead*!" Of course my name is Fabian or in this case: *shithead*.

Every morning she was the first one out of bed and making meals but was always the last one to eat. It was her way to make sure everybody was fed.

We wore decent clothes while she wore the same clothes until they fell apart. She'd never spend any money on herself but only on others. She was always sending stuff to families in need and her grandchildren who didn't live with her. At the time I didn't understand why I was being asked to take goods to other families. Wouldn't we need them later? I had not learned the gift of giving to others.

Grandma's generosity planted those same seeds in my heart. She taught that giving was a reward in itself, never expecting anything in return. She couldn't give me everything I wanted but she did give me everything she had. I couldn't be more appreciative of her teachings although I doubt she realized the impact of her actions. Often we make the biggest impact on others when we don't realize we are doing it.

I took Grandma's advice on studying to heart and became an almost-model student; I still found excuses not to go to school! In fourth grade I told my teacher I was having surgery and wasn't able to attend school; I then told my parents the teacher had surgery and there was no substitute so I had to stay home! (I miss the '80s. No automatic calls home, pesky emails, or texts to rat me out!) After a week Mom didn't buy the substitute excuse anymore and I had to go back but I had a plan. My teacher was too trusting and would send me home if I faked pain from my "surgery". I walked to school and immediately upon seeing my teacher I began my act. I did my best interpretation of an agonizing patient with my classmates who helped me into my seat and walk around. I handed in all my past due homework—I had completed my homework as it was easy for me.

My teacher saw my act of limping while holding my stomach and asked me how I was. I told her I was still hurting from having my appendix removed. She exhaled very compassionately and gently stroked my head and said, "Maybe you should stay home for a while longer until you feel better."

I nodded and decided on the closing move. I moved my hand to the lower part of my abdomen and said quietly, "Yes, Ms. Ferreira. I believe it would be better for me to stay home. My incision still hurts." To add effect I began massaging the "incision". Ms. Ferreira looked confused and then a bit angry, as though realizing she had been played.

"Did your appendix hurt before the doctors removed it?" she asked. I hadn't yet developed the sense yet that would have told me I was about to walk into a trap.

"Yes, Ms. Ferreira. It hurt a lot," I said, still massaging my abdomen.

"Funny," she said smiling, knowing she was about to burst my story wide open. "I know it is much to ask, but could I see the incision?"

I was too smart and had prepared for this. I had covered the "incision" with gauze and tape and sprayed some disinfectant on the gauze, making it look authentic. I lifted my shirt, proud of my preparation for this moment and I saw her smile grow. At that precise moment, I knew I had screwed something up—and I was in a lot of trouble.

She pointed to my "incision" and very calmly repeated "Funny." She continued on, knowing she had won, "I always knew you were an extraordinary student but I never imagined you were a unique human being." At this point I was confused. "The appendix is on the right side of the abdomen, not the left."

Boom. Just like that all my lies were decimated. My face went twenty shades of red and I knew I couldn't talk my way out of this.

"Busted," I said.

She looked at me with infinite compassion as she knew what was coming to me and said, "You have no idea."

I was sent for a visit to the principal's office.

I was suspended from school for a week but when they saw my happiness with the punishment they decided against it, choosing rather to make me stay late. Unfortunately, this also worked in my favor as being home also meant seeing Mom's wrath continually. Ms. Ferreira piled on homework and forced me to read several philosophy and poetry books she thought would bore me to death, but I found them quite interesting and actually learned some interesting stuff that would, later, make me sound really smart in front of my classmates and made more than one girl giggle with my new learned prose.

Lesson of Success

— 5 —

Have all the facts.

Have all the correct facts if you are going to create a story. But, it's more important not to lie to get out of your obligations.

Work smarter so as not to have to come up with stories.

I have built a business that allows me to work whenever I want, from wherever I wish, without the company suffering in my absence. Create a business that is self-sufficient without your expertise. Work hard towards your own reality, molding your future the way you want it.

Lesson of Success

— 6 —

Honest work pays off.

I took Grandma's work ethic very seriously and worked very hard to achieve my goals. One thing I learned in all my ventures is that to be a great leader you should have worn all the hats in your company. Nothing will replace firsthand knowledge of every position, and this will allow you to make better decisions when creating systems and procedures for that position.

SCAN TO MEET MY FAMILY

CHAPTER 4

MY FAMILY—CONTINUATION

Warren Buffett said it best: the difference between a job and a business is if you're making money while you sleep—24 hours a day, 365 days a year—then it is a business. If you have to be there to make it run, then it's a job.

What *you* want is a business, not a job. Why? Because you will never be able to stop working at your job ... until you turn it into a business!

I have been an entrepreneur my entire life, so I'm uncertain if I ever experienced my "entrepreneurial seizure", as Michael Gerber describes it in his book, *The E-Myth*. Rather, I believe I was born with it. It was imprinted in my genes, regardless that no one in my family owned their own business.

Dad was a waiter at a local restaurant and Mom was a janitor before becoming a full-time homemaker. Grandma worked as a housekeeper. Aunt Gladys worked for a lawyer (it was through her that I became fascinated with contracts and the power of words). Everyone around me was employed: they didn't dream of having their own business. The epitome of success was to have a government job— that ensured longevity and a secure paycheck. But I was different. I wanted more than a paycheck.

*****

Dad was a natural at making people feel comfortable and getting them to open up to him. Growing up, I watched a master at work. I didn't realize it at the time (and neither did Dad), but he was a natural at networking and creating long-lasting relationships.

People naturally gravitated towards Dad. He knew everybody and if you needed a favor he was the one to ask, because he would know the person you needed.

He was a handsome man with an Alain Delon air to him; his light blue eyes made the neighborhood ladies blush when he spoke to them. He joked with everyone and everyone liked him. He instinctively knew how far a conversation could go without becoming inappropriate. He simply was a man who never crossed the line.

Dad's job as a waiter at high-end restaurants was the perfect fertile ground for him to make connections. Dignitaries and elites ate at these restaurants while Dad enticed them with his charm. It was as simple as reserving a table so they could be seated immediately. Other times it would be saving the last bottle of that special wine his customers loved so much. He would bring them a dessert that was loved, but not requested.

He had a knack of making his customers look particularly important and relevant in front of those dining with them. And his customers loved him for this. They didn't recognize what was going on, but they loved being aggrandized in front of their significant others, friends, and customers. He did this with every customer.

Dad never went to high school. He never heard of the word networking, much less read a book about customer service. His customer service skills came naturally to him and he honed his skill as the years went by.

In time I would appreciate this valuable skill which he taught me in fostering relationships, which in turn I was able to capitalize on to thrive in the business world. It helped to position my companies as leaders in customer service, treating each customer as the most important customer we had.

I have always been an avid reader. Dad was too. He introduced me to the world of comics and small novels— mostly cowboy stories based in the USA. The comics introduced me to characters who I still think of fondly today. I loved Gilgamesh, an immortal from a time before Christ who went through millennia seeing entire civilizations being built and destroyed; and Nippur, a hero who was incorruptible, mingling with kings and misers without missing a beat. I was a huge fan of these stories and I couldn't get enough.

My best friend's dad, Mr. Maure, worked in a place that sold and exchanged magazines. He supplied us with an endless number of magazines every week. We loved reading them and waiting for new editions was like waiting for presents on Christmas morning. It was pure torture having to wait until Mr. Maure finished reading them so we could borrow them. But it was worth the wait.

Growing up I never went hungry and always had clothes on my back; but it wasn't a glamorous lifestyle. My parents had enough money to feed us and keep us warm but that was it. There was no extra money for vacations or nice trips to the movies and local amusement parks.

The world of reading allowed me to escape reality and travel to exotic places. It allowed me to become a warrior, interact with alien civilizations, become a spy, and—the one that appealed the most—a successful entrepreneur who traveled the world, living the dream life.

Reading taught me many things, opening my mind to ideas I didn't know existed. Comics taught me of valor and courage, joy and sorrow, good and evil; but I wanted more. I was hungry for *knowledge.*

*****

Aunt Gladys lived with my family for as long as I can remember. She never married so her nieces and nephews were her children, a responsibility she took very seriously. Her selfless love for us showed forth in her actions; she seldom spent time and money on herself but always spent it on us younger ones. I thank God every day for her. She was truly my advocate and benefactor, pushing me to the next level: do better and be better. She was smart. She knew the ins and outs of the law, which thoroughly impressed my young mind.

One day Aunt Gladys brought home a typewriter. The office where she worked had upgraded to electronic typewriters and they didn't need the mechanical beasts anymore. I was fascinated with it and loved to sit at the desk, stabbing at each letter as I typed the words that came to mind. Soon I became very proficient at typing, although I had nowhere near the speed and accuracy that Aunt Glady had.

Occasionally I visited her office. I watched her as she typed without looking at the keyboard, simultaneously barking orders to people in another office. She was a typing god. She showed me how much easier it was to use the electronic machine: she could correct mistakes before printing the documents and even use different fonts!

When the bosses weren't there I got to play with the typewriters; it was a great introduction to the nascent world of electronics and computers, tools that in future years would change my life for the better.

My family has been an immense blessing. What they taught me has been invaluable, and what they didn't know, they encouraged me to study and learn. To be successful one needs both everyday life skills and personality as well as knowledge, and my family gave me that.

Lesson of Success

— 7 —

Read. Never stop learning.

It is the best and cheapest education you can give yourself.

Open your eyes and ears and learn from those around you. You could be in the presence of greatness, but if you don't open your eyes to see and your ears to listen, you will never know.

Never pass up an opportunity to read. Even characters in comic books can teach you something if you open your mind.

Always be open and be ready to learn.

CHAPTER 5

THE FIRST PRODUCT

Mom and Grandma would send me to deliver vegetables to our relatives. While my family didn't own a farm, Uncle Francisco did.

Uncle Francisco was an amazing man. He was a hardworking, big-hearted farmer. He always seemed old, which was most likely due to the lifestyle he chose. Day in and day out he worked under the hot Argentinian sun that rose over the Andes mountains, heating the fertility-resistant ground. Uncle Francisco never quit, plowing the hard soil day after day with only a few tools and a conviction that he would make a farm out of that arid land. He succeeded.

At the farm, there was a deep water well which would pump water into large grooves in the ground that distributed water throughout the entire farm, allowing him to grow plump grapes, juicy watermelons, and a large variety of vegetables. His vegetables would one day become the first product I ever sold and allowed me to create my first business.

I loved going to the farm and playing near the well. The main ditch coming out of the well was about three feet wide and several hundred feet long. The water never stopped running through there. Further down the water was distributed by creating dirt dams and redirecting the water that was needed and then creating blockades when they weren't needed. However, the main ditch always held water; sometimes it moved very slowly or stood still, but there was always water. The main ditch had a clay bottom, which felt slimy but made for the perfect natural water slide! My siblings, cousins, and I had a blast playing in that ditch. Who needed a water park? We had endless amounts of summer fun there, and the food was amazing too: fresh cut vegetables, farm raised chickens, and homemade bread. The adults drank plenty of "patero wine" made right there on the farm. Everyone had a really good time there.

Grape harvest was a busy season and I could watch trucks pull up to the farm with their empty trailers and then a short while later leave, trailers heavy with the fresh-cut grapes. Many people came to the farm working from sunup to sundown. The whole operation fascinated me. I always thought Uncle Francisco was rich because of all the people who worked for him.

The people working on the farm didn't look like us. I'd later learn they weren't from Argentina. They came from bordering countries such as Peru and Bolivia to work the grape harvest. It surprised me that people would come from so far away to work at the farm. Uncle Francisco explained that it was really hard work and not many local Argentinians wanted to do it.

I scoffed at Uncle Francisco's words. I had been in the vineyard many times cutting a few grapes to eat, and the work was *not* very difficult. My uncle looked at me with infinite patience and wisdom. He smiled and offered me a job working in the fields. He told me that he would double my salary and that I could stay at his house and eat for free.

I was dumbfounded. He confirmed his offer and took me to the vineyard; introducing me to the crew. The crew had a variety of ages, some young, some middle-aged, some old. A few of the children were not much older than I was.

Later I would learn that they were families and they all worked the harvest, from the grandparents to the grandkids. The older generation greeted me with smiles, while the younger ones looked at me with laughter. They were betting I wouldn't last a day; I would have bet the same had I known what awaited me.

The job was simple enough. Grab a square metal bucket, put it on my shoulder, and then pick a row in the vineyard. With a pair of scissors I would cut a cluster of grapes and put it in the bucket.

An elderly woman in the group advised me to get a hat with a cloth to cover my neck and to wear long pants and long sleeve shirts. I smirked. I wasn't about to dress like them. However, I thanked her for the advice and went back to Uncle Francisco's farmhouse to get some rest in preparation for my start in the vineyard the next day.

I slept well that night and Uncle Francisco woke me up before the sun rose. He gave me a cup of tea with some homemade bread and rushed me to the vineyard. When we got there, the crew was getting the tools and equipment ready for the day.

Grandma Elena, the elderly lady who had advised me the previous day about appropriate attire, took one look at me and shook her head smiling. I was wearing a hat but had opted to wear short sleeves and shorts rather than the long pants and the long sleeved shirt she had advised. I was about to be in a lot of pain for not taking her advice.

A crew member gave me a metal bucket, gloves, snips, assigned me a row, and wished me good luck. Then the nightmare started.

The bucket was far heavier than I initially thought and it was difficult to keep it on my shoulder while trying to cut the grape clusters with my snips. I found a way to balance it all but wasn't making any progress. I placed the bucket on the ground, grabbed a cluster of grapes with one hand and snipped with the other. I would walk a few steps from the vine to the bucket until I cleared the vine. Then I moved the bucket to the next vine. The work was being completed but by the time I cleared the first few vines the rest of the crew was almost halfway finished with their rows. My method was too slow.

I watched the crew, paying close attention to their actions. They balanced the bucket on one shoulder and placed it against the vine and with the other hand they would snip the clusters letting them fall inside the bucket. While the process seemed easy enough it took me almost half a day to somewhat succeed; I dropped my bucket many times but slowly I learned to maintain the balance of the bucket on my shoulder. However, I was still struggling to snip the grapes.

Around 11 a.m. Uncle Francisco stopped by to bring me lunch; I still hadn't filled a bucket of grapes. He sat to talk with the crew and watched me from a distance as I slowly made my way back to where we had started. When I got to the makeshift hut where they were sitting I could see him smiling and talking to Grandma Elena.

Uncle Francisco handed me my lunch and looked at my bucket, which was almost full. He shuffled the grapes around and told me they were too smushed and couldn't be sold at retail stores; they needed to be sent for wine. I looked at him silently questioning why I would care. And then he gave me a small nugget of wisdom.

He said buyers pay less for wine grapes than retail grapes and because of that he paid the crew less. Half to be exact. I was floored. I had worked half a day to make the equivalent of a two-liter soda. That was not going to work for me. I wanted to buy myself some nice sneakers. I needed to pick up the pace. I asked how many buckets the crew had picked. Grandma Elena told me they picked between 25 to 30 buckets. Each. And of the high paying kind. I almost hit the ground crying.

She saw my devastation and took me aside. She began putting some cold rags on my head and neck. She then showed how to balance the bucket and use the snips. She made me practice for a little while and then told me to get ready to complete the second half of the day. She would go with me to my row and continue to teach me.

We went back to work and I spent the remainder of the day fascinated at the dexterity of the old woman and the strength she had to fill bucket after bucket, and then walking them back to the collection point.

With my newly learned techniques I was doing much better than before lunch but wasn't reaching the amount of grapes she was harvesting. At the end of the day she finished with a count of 52 buckets and me … five. I was ecstatic. I was on my way to get some new sneakers!

Uncle Francisco came to the vineyard to pick me up. He gave me some tokens (this was how they keep track of what you have harvested) and took me to the house. He told me to take a shower and then meet him at the dinner table.

The whole day had gone by quickly and I hadn't thought about anything but grapes. But, as I began taking my clothes off the torture began. My skin had been scorched by the hot afternoon sun and the simple touch of my clothes brushing my skin as I took them off sent shocks of pain throughout my body.

I managed to get in the shower but the pain only intensified. Thousands of tiny droplets falling on my sunburnt skin were like sharp needles sending pangs of pain up and down my skin. Soon the pain lessened with the relief of the cold water which made the pain more bearable.

After a long shower I left the bathroom and laid down on my bed to gather my thoughts for a few moments.

I woke up when Uncle Francisco showed up the next morning with tea and bread.

"What happened to dinner?" I was surprised. He smiled and said, "We ate it … last night." With that he walked out of the bedroom laughing.

Still half asleep, I couldn't believe I had missed dinner the night before. I was hungry and devoured my bread, drank my tea, and went after him.

He took me to the vineyard where the crew was once again ready to repeat the labor of the previous day.

Grandma Elena looked at me again. This time she sighed. She rummaged through some bags and gave me an old hat with a dirty piece of clothing hanging from the back.

I looked at it a bit disgusted and she said, "You'll thank me later." I decided to take the hat and followed her to our row.

My whole body was aching and all I wanted to do was go back to the house and sleep. But that wasn't going to happen. I looked back to where Uncle Francisco had dropped me off but he had already left.

Grandma Elena had the bucket up on her shoulder and was ready to get started. I took a deep breath and ordered my legs to move but to no avail. I breathed, trying once again, but with the same result.

My legs refused to comply and my back was starting to have small spasms. I didn't understand what was happening. I tried to speak but my mouth also refused to move; the only noise that came out of my mouth was a small grunt.

Grandma Elena turned around and saw my scared face. She walked back to me and asked what was going on. I couldn't tell her.

She looked at me again and asked if I could move. I shook my head, signaling no. She turned my bucket upside down and made me sit on it.

She headed to the hut and brought a paste which she smeared over my head, face, and neck. She gave me a nasty tasting drink and told me to stay still for a little while.

She checked on me when bringing her loaded buckets to the collection point. I would nod at her and then stay still for a while. Time seemed to move in slow motion. I could see the crew come and go while looking at me and laughing. Grandma Elena would yell at them in a foreign language and they would go back to work.

After a few hours she stopped by again to see how I was doing. She asked me to stand up and poured water over me to wash the paste that covered my skin. I felt a huge sense of relief and to my surprise I could now move ... and talk. I thanked Grandma Elena for the help and asked her what happened to me. She replied, "You didn't listen to the old lady with all the wisdom wrinkles!" She gave me a smile and went back to work.

I went back to work and collected 15 buckets of grapes that day. My body continued aching and I thought I was slowly dying. At the end of the day Uncle Francisco picked me and took me back to the house. This time he fed me and then told me to take a shower—which I did—and once again fell asleep immediately …

My new sneakers were so nice. Shiny. They were white with a red stripe, the same colors as my favorite soccer team. The team called me and asked if I could play for them. I could hear the crowd going wild as I entered the field. The grassy field was soft beneath my new shiny sneakers. The other players looked older than I was. The game started and I started running. Those sneakers were out of this world! They made me run so fast! I could almost fly. Wait … I was flying! I could see the stadium from above, the fans were seated on the grandstands, and the players were getting smaller and smaller as I flew higher and higher. I could feel the breeze and soft droplets of rain touching my face … the rain was cold but so nice. More rain and more rain.

I jolted awake.

I looked around the room and there were no new sneakers, no new stadium, and no sky. But there was a small water leak coming from the ceiling and a few water droplets landed on my bed. I looked outside: it was cloudy but not raining. I was startled. I could see outside; there was daylight already. I overslept! I started to dress and then stopped. I went to my bag of clothes and got some different clothes—some that Grandma Elena would approve of— and got ready.

I started walking down the long road to the vineyard. When I got there, I saw the crew was getting ready to go back to work. It was almost noon. Uncle Francisco was there talking to Grandma Elena; they both were smiling as they watched me approach.

Uncle Francisco called my name and said to me: "I thought you were going to sleep the whole day. I couldn't wake you up so I let you sleep."

Grandma Elena grabbed me by the shoulders and with a smirk said, "Finally decided to listen to your elders?"

Adjusting one of my long sleeves, I replied with a bit of shame, "Yes, ma'am."

That day I had decided to wear long sleeves, long pants, and the hat she gave me, replacing the dirty rag with a clean one to cover my neck. I used some bandages I found in the house to create a padding for my shoulders so I could better balance the bucket on them.

I bent the snips' handles so they better fit my small hands and then wrapped them with tape so they wouldn't slip in my hands. Uncle Francisco studied me, tapped me on the shoulder, smiled, and walked away.

Ready to get going, I grabbed a bucket and went to work. That day I once again completed 15 buckets but it was only half a day. The crew was impressed and they began calling me by my name instead of unpleasant words. I was happy and when Uncle Francisco picked me up at the end of the day I could tell he was proud of me although he didn't mutter those words.

For the remainder of the week I harvested 25 to 30 buckets a day, and while it was still well below the crew's average, I was still a child and I had learned the trade in only a few days.

Lesson of Success

— 8 —

Listen to your elders.

The wisdom they've learned throughout the years can save you a lot of aggravation. You don't have to take their advice word for word, but listen carefully and make your own conclusions. It will save you a lot of time and money not making the same mistakes they made.

I often hear you have to make mistakes to learn from them, but I don't believe that. I believe you will make a lot of new mistakes that you can learn from; there is no need to repeat the ones others have made before you.

Lesson of Success

— 9 —

Focus on improvement.

No matter what area you work in, focus on what you need to do and find ways to do it better. Think outside the box; better yet, don't even think there is a box you need to fit into.

At the farm, there weren't any snips made for kids' hands. They were all big and made of metal but I found a way to make them fit for my hands. Long after I left the farm other kids and some of the women in the crew were using the modified version of the snips. I have made a difference in the lives of others without realizing it at the time (and I was still a child). There is no age limitation for innovation.

You can be brilliant regardless of your age. Your mind is timeless.

Lesson of Success

— 10 —

Never let others decide when you've had enough.

Even when it's out of love. Uncle Francisco allowed me to sleep in on the third morning of work because he loved me and wanted me to rest. However, by doing so he was depriving me of the opportunity of proving I could rise to the challenge. I had no intentions of making a career out of harvesting grapes and he knew this. He always said I needed to focus in school but I wanted to prove I could do the physical labor, and I'm glad I did: these lessons will stay with me for the rest of my life.

Sometimes people—with the best of intentions— will steer you in the wrong direction. Stick to your convictions and remember that no one but you can tell you how far to go.

Lesson of Success

— 11 —

Hard work pays off.

After working in the vineyard for a week I had earned enough money for sneakers, was able to give some money to Mom, and also have some extra to go out to the movies with friends. It was all gone in a short time as I hadn't learned about savings or creating capital, but those lessons would come at a later time in my life.

Lesson of Success

— 12 —

Learn the facts before judging.

During my time at the farm, I learned about the hard life of the migrant workers and their families and learned about discrimination. These lessons would affect my thoughts about immigration and influenced my preconceptions about immigrants.

Immigrants are often misjudged because of race, nationality, and economic status. We can be quick to forget that they are the ones who are actively seeking a better life for themselves and are willing to put in the hard (often manual) labor with long hours to ensure the improvement of their lives.

I saw how other people treated the migrants and accused them of stealing their jobs, but they quickly quieted when Uncle Francisco intervened and offered them the same job. He gave them the same offer he made me: double what he paid the crew if they produced at least the same amount of work. The answer was always the same: no. They'd politely decline (with excuses) and leave. I now know why: it was a very difficult job. The relentless sun of the Andes mountains combined with dry weather carved *indelible* marks in the faces and bodies of the workers.

The crew Uncle Francisco employed were a great help to him and to others like him—otherwise the farmers would see their crops rot due to the lack of employees. I would see this again when I moved to the United States many years later.

There are many aspects of life and difficult situations that we make harsh judgements against before knowing the facts.

Do not judge a person without having walked a mile in their shoes.

Lesson of Success

— 13 —

Discovering new business opportunities.

While working at the farm I learned about many of the vegetables that were grown there, noticing how different they were from the ones my parents bought from the neighborhood stores. I was impressed with the quality of Uncle Francisco's vegetables and not too long after this I had the inspiration to start my very first business: selling vegetables from the farm.

Keep your mind open to the possibilities around you. You never know where inspiration for your "next great big idea" will come from.

CHAPTER 6

MY FIRST BUSINESS

After experiencing the labor-intensive days at the farm, I went back to being a kid. I was broke (having spent my hard-earned money) but I had new shiny sneakers. The shine didn't last very long as I used them daily and soon thereafter they started falling apart.

I asked my parents for new shoes but they really couldn't afford the ones I wanted. Grandma and Aunt Gladys were also overextended in paying the household expenses. I argued that I had torn my sneakers apart doing all the chores they asked me for: from picking up wood for the water heater to delivering all the vegetables Uncle Francisco kept giving to other families. It was their fault my new sneakers were torn, not because I used them every waking minute. My argument didn't go anywhere. And I still had to keep bringing vegetables to others and going to the neighborhood grocery stores to make purchases.

One day I was with Dad at the grocery store where we bought our fruits and vegetables. Dad was bargaining with the shopkeeper over a few garlic bulbs. She was overstating in a loud voice the virtues of her garlic while justifying the price.

Watching the exchange, I thought to myself, "She is nuts if she truly believes her garlic bulbs are that great. A couple weeks ago I took a bunch of those to my relatives that were at least three times the size." Dad and the shopkeeper kept haggling over a price and she kept justifying it by saying they were hard to come by as there was a shortage and they had to be brought from other states.

I shook my head in disbelief, thinking, "That's a bunch of bull. What shortage? Uncle Francisco has lots of them at the farm."

It was there, in between thoughts, and the shopkeeper telling Dad if he could get them better or at a better price somewhere else then she would buy them there too that a light bulb went on in my brain: I started planning what would become my first business.

My business plan was that I would get the garlic from Uncle Francisco and sell it to the shopkeeper. I needed to get in touch with my uncle *now*. Uncle Francisco didn't have a phone (and neither did we) so I had to wait until he came back to my house to ask him.

The next time Uncle Francisco showed up I wasn't at the house but he had left a bunch of garlic. Without telling anyone I grabbed a few and took them to the shop, showing them to the shopkeeper, Ms. Rosita. She looked surprised to have a child trying to sell her garlic but she liked the product.

She asked me how much I charged. I didn't know how to answer her. What did I know about pricing? I told her I wasn't sure but she could tell me. She was hesitant for a moment and then she gave me a price. It sounded like a lot of money but she was speaking of the cost per crate. I didn't know how many bulbs would fit in a crate. She then reversed her decision and told me she would pass on the offer as I was unable to supply garlic consistently. Once her customers would see the garlic bulbs I was offering, then her customers wouldn't want to buy the other bulbs.

Another issue was presented: if she didn't buy the garlic from her supplier where she got the rest of her vegetables, then he might not want to sell her the other vegetables.

I lost my first sale. But I became familiar with something very important in the market. If you have a commodity product that's difficult to find, then you can use it to sell other products that are not uncommon *or* you can dictate the price.

Realizing my garlic tycoon dreams were sinking, I started to panic. I lost my only "customer" and now I didn't know what to do with the "thousands of bulbs I had" (I didn't have any more garlic than what I had in my hand. The thousands of bulbs were a figment of my imagination).

I couldn't understand why Ms. Rosita wouldn't buy the bulbs from me. She said her customers would love the garlic variety and she could even charge more for it. Those last words kept resonating in my head. I knew people loved Uncle Francisco's garlic. They told me that every time I took it to them—then another light bulb went on.

I would sell directly to consumers. Problem solved. I knew how much Ms. Rosita charged for her "inferior product" at her store, so I could sell mine for more and I would even deliver it to people's homes! I was a genius. My garlic empire was back on track.

I still had no garlic to sell as I hadn't spoken with my "supplier" (Uncle Francisco). I stayed close to home for the next few weeks, barely going to the bathroom as I didn't want to miss Uncle Francisco's visit. Then he showed up.

I told him about my plan to sell garlic directly to consumers. He let me talk for a while and finally told me I could have as many garlic bulbs I could harvest and carry them from his farm to my house. For free. I was ecstatic. I couldn't wait to get to the farm to begin.

The next day I discovered my lack of planning.

Early that morning I told Grandma about my plan and she was skeptical. How would I get to the farm? How would I get back? How would I carry the garlic? I hadn't thought of any of that. I decided I would go by bus. She gave me bus fare and off I went in pursuit of my dream.

I had to take two buses to get to the nearest drop off point to the farm, which was a few miles from where the farm was. From there I had to walk. But I was young and super excited about my plan.

I was already spending the money I was going to make from selling garlic: new clothes, shoes, and maybe even a bike! That would be cool!

It took me about 30 minutes to get to the farm, and after speaking to my aunt and uncle they gave me a burlap bag (much like a potato sack) and pointed me in the direction of the garlic patch.

Uncle Francisco gave instructions on how to pull the garlic bulbs from the ground (garlic bulbs are buried).

I walked to the patch and was breathless at the enormity of the task in front of me. There was garlic planted as far as the eye could see. I picked a row close to me and I started pulling.

The work was relatively easy as the ground was somewhat wet and the bulbs didn't offer resistance to being pulled from the ground. I began filling the bag and it was soon filled with about 100 bulbs. I multiplied the bulbs by the unit price, counting my profits. When I finished, I tied the burlap bag closed and headed back to the house.

Walking down the dirt road to Uncle Francisco's house, I dreamed of how I would spend my newfound fortune. Once at the farmhouse, my aunt offered me food and a drink which I quickly consumed. I was antsy to get back to my neighborhood and start making money. Uncle Francisco wasn't home, so he couldn't give me a ride back to the bus stop. I wasn't worried. I told my aunt I would walk and with the bag over my shoulder I did just that.

It wasn't very long before the weight of the bag started to slow me down. I kept moving the bag from shoulder to shoulder, but it got heavier and heavier. I hadn't made much progress towards the bus stop before stopping on the side of the road to rest and then continued my journey.

There was very little shade along the road which made progress difficult. I kept pushing forward, keeping the bus stop in mind. I kept telling myself the bus stop was just a bit further down the road … and then I repeated those words over and over. It took me almost an hour to walk to the bus stop, and when I finally made it I was sweating profusely.

I cleaned myself up as best as I could and sat down to wait for the bus. After about 15 minutes the bus showed up. I paid the fare to the driver and sat in the back. It would take about 30 minutes before we got to the place where I needed to change buses to get home. I closed my eyes for a few moments and when I opened them again, we were arriving at the bus terminal.

I was surrounded by people staring at me with disgusted faces. I wasn't sure what their problem was but since I was soon disembarking from the bus, I didn't pay much attention to them. While waiting for the second bus I discovered why people were avoiding me. The combination of my sweat and the garlic was creating a deep, penetrating stench that could be smelled several feet away. My bus was coming so I decided to proceed towards it. But I wasn't allowed on. The driver on the second bus told me I was not allowed on the bus smelling as I did, so I had to get off.

I didn't know what to do so I went back to the bus terminal bathrooms and washed myself up as best as I could. It helped.

I waited for the next bus to come. I stayed away from the bus stop until the bus was there and then waited until the last person was on board before entering. I paid the fare and went straight to the back.

Luckily the bus was mostly empty and this time I was permitted to stay. Although it only took about 20 minutes, it felt like the longest bus ride of my life. Every person who got on the bus and walked to the back eventually moved towards the front after giving me the stink eye.

I was so embarrassed, but I finally reached home. I dropped the bag in the courtyard and made a run for the bath. It was the only time I voluntarily took a bath and then stayed in the shower long after the hot water ran out! Once bathed, I ate dinner and went to bed, dreaming of all the money I would make the next day.

PERCEIVED VALUE

The next morning, I quickly got out of bed. I skipped breakfast and went to my "bag of riches" and started separating the garlic bulbs from the stems—something I should have done at the farm to fit more bulbs in the bag. Once I was done, I selected the biggest bulbs from the crop, placed them in individual plastic bags and headed out to find customers. My first visit was to relatives as they were close by.

First on my list was Uncle Eusebio. He was a character. Tall, dark, and a full head of white hair. I knocked on the door and he came out. He invited me in and I told him I was there to sell him something he wanted. I showed him the garlic bulbs. He grabbed a bulb and after looking at it he said, "This is one of the biggest I've seen. Thanks!" I was happy he liked it, and then I stretched my hand with an open palm to get my money.

He looked at me with a confused face. "Are you expecting a tip or something?" he asked.

"No tips, just what the garlic costs," I replied and I told him the price. He laughed. I was surprised. Why would he laugh? It was the same price as the grocery store but three times the size. He realized I was being serious and said something I would never forget, "Why would I pay for this when it has always been free? You have been giving me these for free."

Similar situations would transpire throughout the day with every one of my "customers." I was stunned. I spent all day going from house to house and didn't sell a single garlic bulb. I didn't understand. What went wrong with my plan? The words from Uncle Eusebio resonated inside my head, "This has always been free." I couldn't think of a solution to overcome the problem.

I continued to ponder over my problem. Suddenly, I understood. I was trying to sell the garlic to the wrong people. My relatives wouldn't pay for this. Uncle Eusebio said so. They were used to getting it for free; I needed to offer the garlic to those who were paying for it.

The next day I started knocking on neighborhood doors and visiting people I hadn't "seen" before. I did much better. By definition, "much better" was only three customers. I was pleased that I had sold a few, but I also learned an important lesson in sales: rejection is part of the game. Some people would politely decline while some others would yell at me. I didn't like rejection and was upset when I got back home.

Uncle Ricardo was visiting that day. He was a charismatic man; one who I'd work with in the future. But that day he taught me something that I took to heart and which has led to many successes throughout my life: never quit. Use each rejection as fuel to get to your next customer. There will always be people who will reject you but the only person who can determine when you have had enough is you. Don't quit, keep trying. You'll eventually succeed.

Uncle Ricardo worked for a company selling shoes across the state and was accustomed to rejection. I took his advice and the next day I continued with my work, deciding to visit double the number of people I had the day before. That day I closed seven customers. I was much happier but still disappointed it was taking so long to sell my garlic bulbs. I had planned to sell all my stock in one day but here I was making single sales with many people telling me to come back as they didn't have money.

I spoke to Grandma about my dilemma. She was proud of me for trying but couldn't help me.

I went back to speak to Uncle Ricardo and asked him how he sold so many shoes. He said, "I know the type of customer I need for each kind of shoe I have to sell." I was confused so he explained, "I don't try to sell dancing shoes to a construction worker as he might not have any use for them; but he can use some comfortable sneakers or working boots. Also, I make sure I always have something in his price range. Something he can afford."

I knew immediately where the issue was in my sales plan. I was trying to sell a non-essential product to people who could only afford essential items. I was targeting the wrong customers.

The following day I went to a different neighborhood. It was a more affluent area with paved streets and sidewalks and most of the houses had garages. I felt great about the area. I straightened my shirt and off I went going door-to-door selling my garlic bulbs. At the end of the day I had sold to 22 customers. This was triple of the previous day, but still not what I expected. Surprisingly, most rejections weren't about money. They just didn't need garlic now, didn't use it much, or didn't want to deal with the mess and preferred garlic powder.

I thought about this and came up with one of my best ideas yet. I would peel the garlic and pack the clean cloves in a small clear bag with a piece of paper containing tips on ways to use garlic. I went to work cleaning the bulbs and separating the cloves. I hand wrote approximately 30 tip sheets. Writing the tip sheets took longer than expected, as I had to get ideas from Grandma and Mom, and then prepare sentences that made sense and make it all fit on a small piece of paper.

My calligraphy was horrendous and took me a long, long time to make the tip sheets look readable. The next day I was ready with 30 packs and headed out with the goal to sell them all by the end of the day. On my first attempt I sold two!

The customer told me it was a great idea and she loved that they were the same price as a garlic bulb at the store. That sent alarms ringing in my brain. I immediately realized I was leaving money on the table. My bags contained about three times as much value: I was underpricing them. I decided to double the price and see how it went. In the next two hours I'd sell every one of the bags I had prepared. I had successfully started my first business and learned a critical lesson about pricing: **The right price is not always the lowest price, but the one where customers can perceive the value of the proposition.**

I went home, and with the remaining bulbs I prepared more bags for the following day. Once again the most painful part was writing the notes.

The next day I sold out in less than two hours. I had sold all my inventory of garlic in three days by realizing who my customers were and what they wanted to buy. I didn't know it at the time, but that was exactly what I had done. I had identified who the right customers for my product were, and thus I could stop wasting my time pursuing the wrong customers. Also, I had packaged and priced my product in such a way that the perceived value of the offer was much greater than the price the customer was paying for it, which made for easier sales.

This principle would serve me well in the future and would save me plenty of aggravation by targeting the right customers for the products or services I'd want to sell.

I also learned that prices were not set by me but by the expectation of value by the customer. In other words, it didn't matter how much I thought my product was worth; it only mattered how much value the customer perceived they were getting. The higher the perceived value, the higher the price I could charge.

Without inventory (garlic bulbs) to sell and knowing I could sell plenty more than I expected, my challenge resided on how to scale "business operations." I needed more garlic—which I knew I could get from my uncle's farm—but I also needed a more efficient way to print the notes, in addition to reaching more customers in the same amount of time.

Like most businesses, I went to look for help. I went to my friends and asked them, *not* if they would want to work, but if they wanted to make some money. They were skeptical at first but I offered them 25% of the sales. I gave them an approximate number based on what I had sold and they were very on board.

We arranged to go to Uncle Francisco's farm the next day so we could pick garlic bulbs together.

My idea was to get as many bulbs as we could, this time without the stems, so we could fit more in the bags. We would come back, clean them, separate the cloves, and then all of us would write the tip sheets so we could have bags ready to sell. When we reached the farm Uncle Francisco was there and I told him how well the first round went. He was very happy. I offered to cut him into the deal but he wouldn't hear anything about it.

Uncle Francisco wanted me to do well. He took us to the garlic patch and waited for us to fill our bags plus a few other bags he had brought with him. We went back to the house, removed all the stems and finished with four fully packed bags which we could barely lift. Uncle Francisco offered to bring us home in his blue Jeep, which was an immense blessing.

When we returned home, Mom was upset that we had imposed on Uncle Francisco for the ride, but he said he was happy to do it and it gave him the opportunity to bring her vegetables. We let them be and got to work.

I knew I needed a system for efficiency and to set things up so two people cleaned the garlic bulbs and broke up the cloves while the third person wrote the tip sheets. I got a few more friends to help write the sheets in exchange for promises that we would take them out for ice cream when we sold all the packages. I discovered two of my neighborhood friends had good calligraphy and were quick writers, so I promised them I would also buy them ice cream if they helped. Before I knew it I had six people working on the enterprise.

It took us the remaining part of the day and all of the following day to prepare all the bags. When we were ready we had hundreds of bags. Uncle Francisco promised us to come back in a few days with another load of garlic so that we didn't have to go pick them up, so all we had to do was make our supply last for a few days.

With my two friends, Gustavo and Felix, we headed out to sell our goods. Felix was older than Gustavo and I and as a result he was much stronger than we were. Gustavo and I made Felix carry most of the weight. He also wasn't very good at talking to people, so we decided that Gustavo and I would go from house to house. Gustavo would take one side of the street and I took the other one. Felix would carry our stock down the block and wait for us at the end of the street before moving to the next street.

By the end of the day we had sold all of our stock. Every single bag. We were ecstatic. We couldn't contain our happiness. We counted all our money and split as agreed. I kept 75% and they received their 25%. That quarter was more money than they would sometimes see in months. We couldn't believe we had made that much money in one day!

They wanted to go spend their money, but I wanted to go home—and plan for more days. It was more important to me that I keep making money rather than to have made a bit, but for my friends, it was all about enjoying the fruits of their hard work. They persisted in enjoying their hard-earned money; we went downtown to play videogames and eat pizza.

The thought that tomorrow I wouldn't have anything to sell and that Uncle Francisco wouldn't be back for a few more days spun through my head. My friends spent most of their money while I kept most of mine, thinking I'd have to buy more bags, pay the other friends who helped, and give some to Mom.

By mid-morning the next day, I had spent most of my money buying supplies—bags, papers, pens, crayons, and peelers. I bought a few candy bags to keep as currency and gave Mom some money. She was surprised I had made that much. Mom said she would get one of my aunts to tell Uncle Francisco that we needed more garlic ASAP.

Meanwhile I hired three kids to write tip sheets and get the bags ready for the garlic. In two days I would get over one thousand bags ready but with no garlic.

Uncle Francisco arrived on the third day with his Jeep loaded with garlic. It was more than enough to fill several thousand bags. Mom's courtyard turned into a mini factory with supplies in one corner, produce in the other, and people (my friends) scurrying from one place to another.

Throughout that time, I learned giving my friends 25% of the sales was too much as I had to take care of all the expenses. Once I determined the expenses, I called them to have a meeting, explained the figures, and told them I would still give them their 25% but after expenses (gross profits). They agreed. We also agreed to pay the other kids in money rather than candy and ice cream. This taught me the importance of knowing the numbers of my business and the correct terminology. Sales didn't equal profits.

We had a fully functioning business now. The "partners" would go out to sell while the other kids would stay behind cleaning and packing garlic. It was great. Every day we sold everything we prepared and Uncle Francisco came every week to replenish our stock. We were working at full capacity. But we soon started running into growing pains.

Our main concern was that we were running out of affluent neighborhoods around us and as we didn't have transportation—that was a big issue. At the rate we were selling and completing neighborhoods we would run out of customers within a week, and we had only started the business three weeks earlier. Garlic can sit unused for months so we didn't have demand for repeat customers. I was devastated. My "empire" was crumbling and I didn't know what to do.

Mom came up with an answer: shut the business down. I looked at her like she had escaped from a psychiatric hospital. As calm as I could I asked why she even thought

I could do that. She smiled and said, "School starts again in two weeks, you might as well enjoy the rest of the summer before you go back to studying." The nerve of that woman. To think that I would shut down my business because of something as trivial as school. It wasn't like I was 11 … I was already 12!

Regrettably, nefarious forces would conspire in the upcoming days (Grandma and Uncle Francisco) and they would shut down our supplies while another confabulator (Mom) would evict us from our leasehold (her courtyard), effectively putting us out of business. I was disappointed, but while paying my partners and helpers I realized I had made more money in less than a month than Dad had over the entire summer.

I had become a neighborhood legend; my friends couldn't stop talking about how much money they made with me and over the following months people would approach me asking for a job. I was able to get new clothing, some sneakers, give a good chunk to Mom, and still have enough leftovers to last me for months.

Lesson of Success

— 14 —

Always get paid.

Don't give products or services away for free or people tend to think what you do or sell has no value. This is especially true for your services.

In the construction industry the concept that contractors should do estimates for free (as if their time, expertise,

---

and other resources needed to produce the estimate would have cost nothing to acquire) is prevalent. The industry has used free estimates as a way to get leads for work; however, the issue is the leads obtained by giving your services for free are most likely not good for you.

The sooner you learn who your customers are, the sooner you can come up with added value services that your customers will perceive as a good deal. Even if that service is just estimating, there is no reason why you need to do that for free. While many contractors would do that, you don't have to.

Find a way to offer something that will resonate with the people you want as customers and stick to it. You will get more qualified leads and close more deals than giving stuff away for free.

Lesson of Success
— 15 —

Price your products and services appropriately.

Pricing is never based on how much you would pay for something; it is based on how much value your customers perceive the product or service has, which will determine how much they are willing to pay for it.

Lesson of Success
— 16 —

Plan to succeed.

Don't only make contingent plans in the event something doesn't go the way you planned. Many people only make plans in the event things go wrong. Make plans to succeed so you can be ready to expand.

Lesson of Success

— 17 —

Compensate your people adequately.

This could be more than money, but ensure you do something that's fair and enticing to them. You need people to expand your business. Otherwise, you are limited by what you can do and you'll never be able to expand or to have a proper enterprise.

If you do everything yourself, you just have a job. Don't be afraid to surround yourself with talented people, but don't see them as competition, even if some of them might become one.

People who leave your business to run their own business can help you keep growing. You need to find a compensation formula that works for everybody involved.

Lesson of Success

— 18 —

Businesses are much more complex than childhood enterprises.

Pay close attention to details: regulations, laws, licensing, and taxes. I applaud you for wanting to start your own business, but don't enlarge the statistics of failed businesses for lack of preparation.

Learn what you need before launching.

Prepare to succeed.

CHAPTER 7

GROWING UP

I would go on to have many other jobs before I reached my 18th birthday.

I was a helper baking and distributing bread with my uncle Ruben. This was a hard job. We would start really early in the morning, around 4 a.m. or earlier, and we would have to prep the dough, knead it, bake it and then go out to sell the bread. We would park an old truck as close to the delivery places as we could, but most days, we would have to walk long distances with large baskets with bread.

I worked with my uncle Ricardo selling shoes. We would drive long distances to different parts of the state and sometimes outside the state to deliver shoes. This was fun. I had long talks with my uncle about many things. He was fun to hang with. Later on in life, I would work again with him, running a convenience store of sorts that he had in a pool hall, and where I would become a great pool player from practicing long hours.

I worked with my aunt Rita and cousin Andrea selling stuffed dolls in fairs, and I was one of the best salespeople my aunt had. I loved sales. I could sell anything. Sometimes we would set up next to the people selling toys, and I'd sell all our stuffed dolls by preaching to parents the benefits of letting their kids go back to simpler times, when we were free of so many manufactured and imported toys. If I could have seen my future children and myself 30 years in the future, addicted to electronics and the Internet, I'd probably have kicked myself.

Then, I worked in a bowling alley with my friend Felix. Here the job was to set the pins in place after people played, so you had to stay very alert to avoid pins or balls hitting you. As you can guess, over 30 years ago nobody was too concerned with workers' safety.

I worked in a quarry, also with my friend Felix, where we had to break large stones with even larger sledgehammers, kind of like the prisoners' work you see in the movies, but paid. The pay was not much better, but at least, the work was voluntary.

I also worked as a construction laborer digging ditches and foundations. It was brutal work and incentivized me to find something where I could use my brains more than my muscles.

I got a job working as an errand boy at a local pharmacy; I quickly learned about different medications and compounds and offered to be a door-to-door salesman. Many of my customers recognized me from my garlic business and were happy to order medications through me. Here I learned the importance of happy customers: they were much cheaper and took less time than attempting to acquire new ones. I greatly increased the pharmacy's outside sales.

Before I turned 18, I had worked in several different lines of work. All hard work. But it was also the age that I began hanging out with the wrong crowd.

Concerned about the path I was going down, Aunt Gladys intervened by giving me a library card: this gift changed my life.

The library card gave access to a private library of thousands of mostly new books. A good number of them were on computers. The library had volumes of books from the greatest authors. Week after week I devoured the books, opening my mind to a world of knowledge far beyond my years.

I would retreat to a nearby premises with a homeless shoe shiner who had set up there. Mario's appearance was unpleasant and there were obvious signs he liked showering even less than I did, but he was well read and I enjoyed the conversations we had. He read hundreds of books and could speak with authority on many subjects.

The librarian would allow Mario to borrow books, which he handled with great care and precision. We could talk for hours: politics, economy, businesses, philosophy, poetry, and science.

Mario never told me what he did before becoming a shoe shiner, but I could tell there was an interesting story beneath the surface of what he did reveal. I had many theories about Mario but I never was able to confirm any.

Mario was a very private man, never speaking about himself or his family, but he would listen to me and offer advice about mine. He was a unique individual who taught me many valuable life lessons.

My interactions with Mario gave a foundation and understanding about creating relationships with people in vastly different economic conditions—lessons that would lead me to close some of my best business deals.

Sometimes the people you least expect help you connect with those you need to be in touch with. Never dismiss people based on their professions: a janitor has the keys to the CEO's office—and perhaps also his ear.

Sadly, one day Mario never came back to the library and we never knew what happened to him. I think fondly of him and will always be grateful for the life lessons he imparted.

The enormous amount of knowledge I acquired by reading books would help me do extremely well in school as well in life.

Lesson of Success

— 19 —

Never judge a book by its cover.

We never have an opportunity to learn from other people if we set them aside because of the way they look or what they do.

Exercise empathy with others, not judgment. Remember you can't really judge other people without having walked a mile in their shoes—and you probably don't want to do that—so reserve judgment.

CHAPTER 8

HIGH SCHOOL—A MULTINATIONAL ENTERPRISE IS BORN

In high school I was well known by the faculty as many were my clients. A friend and I started selling clothing imported from a bordering country where it was made much more cheaply. I made a decent commission but quickly realized I could do the same by myself and enhance the business by bringing goods from Argentina to Chile, hence creating my first multinational business. I could sell the clothing in Chile while purchasing more clothing to bring back to Argentina.

I was a very good salesperson and from an early age I understood that customer service would differentiate my products from that of the competition. Every product or service I sold would come with a satisfaction guarantee. I went as far as replacing damaged goods with either a replacement item or with store credit: both were new concepts in the area which no one used. I was able to provide this service very effectively as I had profit margins large enough to cover the damaged and defective products. No profits result in no customer service, as one cannot afford to provide it.

My clothing business was a lucrative deal. I took leather jackets from Argentina across into Chile, selling them for double what I had paid. I would then use that money to buy clothing—sports clothing, shirts, jeans—and bring them back to Argentina. I'd sell the clothing using a four-payment plan: the first installment covered the cost and a small profit, and the last three payments were strictly profit. I had fantastic 80% margins which allowed me to offer the financing and satisfaction guarantee.

My clothing made high school great. I was making good money selling the clothing and I had easy access to potential customers. Most of the faculty—including office staff, teachers, and janitors—would purchase my clothing, and as they were frequently indebted to me, I was extended certain courtesies normally reserved for faculty or staff. I learned how discretion got me much further than flaunting my "connections."

My homeroom class and teachers had the best access to the clothing. My teachers encouraged my entrepreneurial dreams and frequently accommodated my schedule, giving me permission to reschedule tests and exams as I was often absent due to traveling. Occasionally, they would allow the test to be taken immediately after the lesson was explained.

Learning came naturally to me. Unlike most of my classmates I could master subjects in a very short period of time. I found subjects such as science, literature, and philosophy were extremely easy as I was so well versed in them thanks to all my reading. Although I didn't read much about math, that too came easy. There was one subject, however, that I did not excel in: art. I was terrible at it. My drawings reflected the skills of a five-year-old (and those were my best drawings). It amazed my teachers that I could excel in other subjects but be so awful in the arts. To this day, I believe they passed me because they felt sorry for my lack of skill.

Over time I'd become a patron of the arts supporting artists from around the world in varied disciplines: from painting to sculptures to music. I knew firsthand how difficult their professions were; I had and still have enormous respect for artists.

During my high school years, I was in charge of anything that funds needed to be raised for, from lab supplies to senior class trips. I spearheaded all the campaigns. I was a born leader, never afraid to take on new challenges. My classmates recognized this and allowed me to lead. As a result, I was the class president and on many fundraising committees.

My clothing empire collapsed when the cost of leather goods skyrocketed in Argentina. Border restrictions further impeded operations, leading to the closure of my first multinational business. Shutting the business was a hard blow to my self-confidence but also taught me valuable lessons.

Lesson of Success

— 20 —

Every other business provides good customer service.

Every customer also expects good service so the goal in your business should be to provide an *experience beyond your customers' expectations*. In Argentina return policies or warranties in the clothing business were unheard of. I introduced the idea, but only because I could afford it.

Don't give your products or services away; price them according to the benefits your customer receives when they purchase them. Sell on value, never on price.

Lesson of Success
— 21 —

Be prepared for extreme situations.

New government regulations would shut down my clothing business. The recent global pandemic shut down millions of businesses, thereby shattering lives.

You need to be prepared to pivot, find new opportunities, and to keep moving forward. There will always be challenges. Be prepared to conquer the challenges.

CHAPTER 9

COMPUTER BUSINESS

Right before I started high school Grandma insisted I learn about computers. I wasn't thrilled with the thought, but agreed: I'd do anything for Grandma. It was her way of keeping me busy and away from my neighborhood friends, most of whom would eventually end up having issues with the law. Learning about computers had the added bonus that the skills acquired could help me in the future to land a good job. Despite her age and generation, Grandma had the good sense to anticipate computers would be the future. She pushed me hard to learn how to use them, and eventually they would become a skill. That skill made me a very comfortable living for many years and also allowed me to design the construction software that would eventually become the largest venture we ever created.

I pivoted my ambitions and studies towards computers and learned everything about them, from assembling to programming. I discovered I had a gift for understanding complex equations and formulas so I became very proficient at programming them.

Towards the end of my high school years, I met the girl who would become my wife, my partner in life, the wind in my sails, the mother of my children, my everything. Agata was the only person who never doubted that I'd succeed in life. She didn't realize how much she would be a huge part of my success, but her faith in me was steady, unbreakable, and unconditional. She has been my inspiration from day one and to this day is my fiercest and staunchest ally.

I loved money, and school was in the way of me realizing my earning potential, so I dropped out of high school in my last year. Years later I realized I wanted a formal education and went back to finish high school and then onto college. I focused on making money. I was a hustler always eager for the next opportunity. Opportunities came, but I lacked focus. (Much later in life I'd realize having a laser focus would allow me to achieve incredible success in life.) Emotionally it was a confusing time as I kept learning computer science but would jump from job to job trying to make ends meet. To add to the challenges in life, I was presented with an unexpected gift: my oldest daughter, Alexa.

Alexa's arrival brought conflicted feelings to my life, but she also brought clarity. From the moment I held her in my arms, I knew she would change my life forever. And she did. Her presence in my life was the constant reminder I needed to do better in providing for her.

Over the next two years I became a computer entrepreneur, opening my first formal office from where I'd operate my first official company: Lions Computers.

At Lions Computers I learned daily about entrepreneurship and business practices such as branding, customer service, and networking. These helped me close better and more deals than hustling.

Let's take a minute to cover these points.

Branding

We hear much about branding but what is it? Branding is knowing your vision and your goals and infusing every part of your business with it. Actions, mailings, and marketing materials—from the business cards to a website to national TV advertising, and even to franchising—should be saturated with your message.

Customer Service

I have spoken of the importance of customer service from my time selling garlic to my clothing enterprise. When you earn enough to make a reasonable profit, you have the resources to ensure that your customer service is top-notch, and you can afford to be customer-centric, irrespective of the cost to the company. Just remember: having no profits means you cannot afford to provide customer service.

We speak about customer service with employees and customers, but actions always speak louder than words. A prime example of excellent customer service is from Nordstrom's, a major national department store.

For many years all department stores had a 100% return policy, although it did not always meet compliance by the staff or management. Nordstrom's had this same policy, even going so far as to do a national advertising campaign after the following incident.

A customer went to a Nordstrom's store complaining about his new tires. He had them removed and brought them to the returns desk in a shopping cart, telling the young man behind the counter that he wanted a refund. The young man (a bit flustered) made a few brief phone calls and determined the tires retailed for $295. He gave the customer $295.

What was so unusual about this transaction? Nordstrom's never sold tires! The young clerk had called a tire store for the price of tires, gave the refund, and then called management to find out what to do with the tires. When management discovered the clerk's *faux pas* they turned it into a national advertising campaign about real customer service. This is what I call customer service. Why stop at providing good service when you can provide your customers with an experience like none other?

Networking

There is a popular saying, "People buy from those who they know and trust." Well, it's true. Customers need to *know* about you—through advertising or referrals—and they need to *trust* you before buying from you. It's difficult to reach people with both of these aspects without investing a significant amount of money in marketing and advertising, but it becomes much easier when you network.

Being part of a networking group is like having an outsourced sales force. You concentrate on knowing each member of the group and educating them about yourself, your company, and your products or services.

The group will be on the lookout for business opportunities and when they find one, they ensure contact will be made between both parties. When the potential client reaches out, a certain level of trust has already been established, as they have been referred to you by somebody they know and trust.

Make a point of becoming part of some networking organizations. If you're committed to the group, it will help bring success to your business.

*****

I had a knack for computers and spent hours learning everything I could about them. In time I became an expert in computer hardware and software programs. I helped people with their computer issues; computers were a new industry with few skilled experts. I was very good with them and word quickly spread. I was receiving calls even from government officials and major corporations requesting technical help.

Even when I was 19, I looked 25, so people never questioned my age and I never mentioned it either. Business quickly grew and I was making good money but had no time to spend with my family or friends. That's when I realized I didn't have a successful business but rather a busy job. Many entrepreneurs fall into the same trap. They quickly get busy and never work in ways to transform their job into a business.

I often get this question: how do you know when you have a business and not a job? My answer is simple: if you're making money when you are not working, then you have a business.

My problem with the computer business was that while I was doing well financially and had good customer retention, I was not making money when I wasn't working, thereby creating a need to work all the time. I never took the time to develop my business into something that didn't require my skills; the entire operation depended on what I could produce and how many customers I could serve.

Internet access was very limited in the late '80s and into the '90s, and access to information was largely limited to books. There was a lack of business mentors, and there were no TV "gurus" mass-producing infomercials selling the latest book or course that would change your life dramatically. I had to learn by trial and error.

Insight One

My first insight in the computer business was realizing I could only do so much, so I hired my best friend, Gustavo.

Gustavo is a fiercely loyal man who would follow you to hell, if you asked him to. He is short—but his temper is even shorter: sometimes he would speak in a low voice or just growl. He wasn't the friendliest person but he always watched my back.

I taught Gustavo how to work with computers. He learned quickly and with his help I expanded Lions Computers. The profits were very good. Regrettably, I didn't gain more time to spend with my family and friends. I now needed to manage him in addition to my regular customers.

Insight Two

Soon after hiring Gustavo, I discovered that I needed to empower my employees to make their own decisions.

A mistake I made when I hired Gustavo was that I taught him how to work with computers but I didn't empower him to make decisions, nor did I incentivize him to do so. Much less, I didn't teach him that it was fine to make mistakes so long as we learned from them.

When an employee makes a decision that results in an error (or has negative consequences), it is better to use these as teaching moments on your way to independence rather than to create employees who constantly need instructions, answers, information, advice, or decisions. From the start, teach your employees the skills they need to succeed in their role, and let them go to work: most likely they won't burn the building down or give away the cash register.

So now, with Gustavo working with me and an increase in customers, the work kept piling up as work wasn't being completed without my instruction.

Once I realized this, I was able to fix the problem by explicitly empowering him to make decisions without fear of retribution, thereby incentivizing him to provide excellent customer service (which forced him to smile from time to time!) and to share his mistakes with me, so I could learn from them as well.

Insight Three

With Gustavo now fully trained I had some free time, which I used to create procedures to manage workloads and customers' requests and which, in turn, helped us to be more efficient, productive, and profitable. This third insight proved to be crucial to the business—and will be the same for every business: invest time into creating systems and procedures that allow your organization to be run systematically and profitably by anybody with only a minimal amount of training.

My first round of systems and procedures were very complex, and while I could navigate them perfectly (as I had created them), Gustavo struggled with them, which slowed him down (and made him crankier than usual). Once I worked with him to create new ones that simplified the flow of work, things improved. In the end, we served more customers in the same amount of time which increased our profits and our job satisfaction—I was happy with the results and Gustavo had a strong sense of accomplishment.

*****

Business was great but competition was increasing at a rapid rate. Our profits began to dwindle which made me think if I lowered my prices I would get more work. And I did. Work began to increase but it was mostly unprofitable work; this kept Gustavo happy because he was busy fixing computers, but it made me extremely unhappy because I was making no money.

Insight Four

The lack of profit also turned into a loss of customer satisfaction which gave me a fourth insight: you cannot make your customers happy if you are not profitable, as you just can't afford to please them.

With impending doom looming on the horizon as our cash flow dipped lower and lower, I spent many days trying to find a solution. I raised prices and increased our profits, but customers weren't any happier and we lost referral business and returning customers. I couldn't charge more for the same crappy service when the competition was charging less, and I was still providing the same service as everybody else. Customers didn't have a problem paying more, but they wanted more too.

I thought about the customers we had or wanted. I thought about what their problems were and what their needs were. I thought about what we were doing to reduce those pain points. I learned about what my competition was doing. Finally, I spent time talking to my customers, which taught me a lot about their needs and wants, and how I could meet those desires.

I finally came up with a plan: we would raise our prices but would provide a service experience so great that our customers would beg us to charge them more. (I was young and had high expectations!) The service plan worked but customers never begged us to charge them more! However, we quickly became the go-to company for IT services. I spent money on additional computers and provided temporary replacements to our customers while theirs were being serviced. We then did the same for printers, which ultimately led us to open a new line of business: selling consumables.

The competition eventually caught onto what we were doing, but while they could deliver a similar service they couldn't deliver it the same way we did. By the time they improved their customer service, we had already improved our service again.

Insight Five

From a very young age I learned one of the most important lessons in business: never compete on price. The computer business only enhances that principle. Lowering prices to remain competitive is a race to the bottom. Most people's first instinct is always to sell cheap. I always prefer to sell less with better service and higher margins. In the years that followed, I realized many businesses fail because of falling into this trap.

There is a real-life story that shows the consequences of the race to the bottom. Two grocery stores across the street from each other were competing on price. All their advertising—flyers, store posters, etc.—indicated their prices were unbeatable. One day the manager of Store B decided to drop the price of toilet paper from a dime a roll to nine cents a roll. Store A couldn't deal with it and lowered their price to eight cents a roll, and down the prices went until Store A was at two cents a roll. Then, Store B put up a sign: one cent per roll! That sign … was directly beside their GOING OUT OF BUSINESS signs.

Pricing your product or service cannot be left to others. You cannot price your products and services based only on your competition's prices. While you need to be competitive you also need to choose at what level you wish to compete. You don't have to compete at the same level your cheaper competitors are. In the same way you can play baseball at different levels, Pee-Wee, recreational, minor, and majors, you can also decide where you want to compete with your products and services. And if the league where you want to compete doesn't exist, that's great; create your own. But whatever you do, never compete on price.

Innovating

We created many innovations during our time at Lions Computers. We quickly understood how computers were changing the world and that customers had an increasing need to learn more about them.

Many parents saw the same patterns in the industry and desired their kids to learn how to use computers. Many families couldn't afford the cost of a personal computer, and for many it was also impractical to send their kids to specialized places to learn about them.

To fill that need we created a program that provided a basic computer which we would install in their home. We provided training for the entire household to learn how to use it. We rolled the cost of the computer and the training into one packaged bundle and spread the cost over installments throughout the length of the training. At the end of the program, the customer got to keep the computer.

The program was a smashing success. We struggled to keep up with demand. While the program was very profitable as I covered the hardware costs with the first payment, I needed capital to finance the purchase of additional computers and to cover the costs associated with getting the program up and running, to the point where the cash flow would support the business.

Another insight was learned during this time: you can be as easily crushed by success as by failure.

We got so many orders we fought to keep up with demand. Eventually we couldn't keep up anymore and started to have issues delivering equipment because we couldn't afford to buy it in bulk quantities. Through the contacts I had made over time, I found my first investors. Using my investors' money I traveled to Buenos Aires, Argentina's capital city, to purchase computers in bulk and send them back home. However, I couldn't buy them fast enough, so I spent a lot of time and money traveling the area to find equipment.

Business was great, but when we originally priced the program, we didn't contemplate all the additional expenses for traveling and business overhead, nor that we would be using money from investors who wanted a healthy return on their capital (greedy bastards!). We started losing money faster than we could make it, and before we knew it, we were closing the doors to Lions Computers and lost the investors' money.

The experience with Lions Computers was a painful one. Not only had I lost my business but I had failed the people who invested in me. I made the stupid decision to hide from my problems, something my creditors didn't appreciate, and which didn't end well.

Over time I learned that running from your problems doesn't solve them, but only postpones the inevitable. Problems and creditors will eventually catch up and the longer you try to postpone facing the problems the larger the problems will manifest themselves. It's much better to tackle the issues quickly because the sooner you do, the sooner you can start over and make things right.

After some soul searching and after confronting my problems, I decided to pick up the pieces and start over.

I had a lot of IT knowledge and many happy customers, so I went back into the IT world and began working as a consultant. Since I didn't have a storefront, it was hard to inspire confidence in people to hand over their equipment for repairs.

I tried to find employment in established computer stores but either they couldn't hire me or didn't want to add to their expenses. Ultimately, I reached an agreement with a couple of businesses. I brought them my customers' equipment, which I would repair on their premises in exchange for sharing the profits.

I ended up working mostly with one medium-sized company whose owner offered me the best profit split. I was a very good salesperson and kept him happy with a steady volume of business. He attempted to hire me as an employee, which I declined, as I trusted my commission from sales would allow me to make much more money than I would as a salaried employee.

The first computer viruses began circulating and I became fascinated with them. There was antivirus software available which we would use to clean the computers, but occasionally we were unsuccessful, resulting in lost information and customer frustrations.

As viruses became more prevalent, I kept insisting on investing in better tools to combat them, but the owner didn't listen. I invested a significant portion of my money in getting the latest tools to help my clients, but it was a losing battle. New viruses were being developed and we couldn't keep up, and without the proper tools many customers were looking for somewhere else to go for better service.

I decided to specialize in computer viruses. I learned how to program them and then to deactivate them. I spent time researching and learning everything I could about them until I became an expert on computer viruses. I was making a decent living until Michelangelo came into existence. Michelangelo was a computer virus that infected DOS operating systems computers (the vast majority of computers at the time were operating in DOS). On March 6th 2021, the virus caused users to "lose" their data when they booted their computers.

In all reality, the data was never lost but was inaccessible to the average user. I discovered the problem and then quickly developed a way to recover the "lost" data. I became an overnight sensation. I was the only person in town who could recover the data for hundreds of users: lawyers, doctors, government officials, etc.

Suddenly my services were once again in high demand because I had a skill nobody else had: the ability to give people back their data. The owner of the store where I operated was extremely happy with the influx of new business but I wasn't, as our arrangement had remained the same.

I renegotiated. He had no choice, but I had proposed a fair deal slanted in my favor as I was the one who was bringing in the money. The arrangement went well and I was soon offering seminars for companies on computer viruses.

Throughout the experience I learned another lesson: become an expert in your field. No matter what field it is you must invest in yourself and never stop learning. *Never.*

I would go on to have several other businesses before I would make my way into construction: a computer store, hardware distributor, software programmer, online shopping, and others.

Lesson of Success
— 22 —

Focus.

You will achieve much more in life by focusing on one thing at the time. Pick one thing, and then plan, execute, correct, and repeat. No matter how small the task or goal you choose, get it done; then move on to the next.

Lesson of Success
— 23 —

Know why you need employees.

Before hiring your first employee, think carefully about what tasks this new person will help you accomplish and how that will produce more time for you to focus on other aspects of the business.

Never hire new employees because you assume you need them. You need to have specific tasks and those tasks must be money-producing tasks and not just busywork.

Lesson of Success
— 24 —

Profits.

Put profits at the front and center of your business. Don't be ashamed of them. Profits and money are not dirty words.

Be proud of being a profitable business—your business can't function if you don't have profits.

Lesson of Success
— 25 —

Aim to be the best.

No matter what you want to be in life, aim to be the best of the best. Wherever you end up, aiming for second best is not an option.

Lesson of Success
— 26 —

Prepare for success, again.

I spoke about this before, but let me repeat it once again. Many businesses fail because they don't have plans if the best outcome happens—they are crushed by their own success. Don't make this mistake. Plans for undesired outcomes should always be secondary to plans for success.

Lesson of Success

— 27 —

Don't run away from your problems.

Problems will always find you. Face the music quickly and swiftly so you can start on the road to recovery sooner. The faster you face your problems, the faster you can move on.

CHAPTER 10

MOVING TO THE USA

At the end of 1999 I received a job offer to work in the USA. The economic situation in Argentina was rapidly deteriorating and once again I found myself struggling to survive. A few friends and I had started selling a computer-based English language course and although it did great and sold well, sales started dwindling because of the economy.

The creators of the software lived in Miami. We spoke several times about their contacts with large companies, such as AOL, and how they could help me get a job with AOL. I flew to New York for a job interview. I was offered the job but with that came a difficult situation: I had to choose between Argentina and the USA. I decided I had to move. It was time for me to leave the nest and fly away.

My wife, Agata, and I decided to make the move with our daughters Alexa and Karen (who had arrived in our lives five years earlier). There were a lot of preparations to be made: selling our few possessions, getting our visas, and saying goodbye to family and friends.

To ensure a smooth transition from Argentina to the USA, Agata and I determined that I would go first and then she and the girls would join me a few months later once I was settled and all our paperwork was completed. And just like that, I left. Without realizing how much my life—and my family's legacy—was going to change forever, I grabbed a suitcase, stuffed it with a few belongings, said goodbye to my family, and jumped on a plane bound for the USA.

As I took my seat in the airplane I reflected on the year and months leading up to this moment. I felt free. Before the plane touched down in the land of the free and the home of the brave, I felt liberated from my past failures, my fleeting successes, and was ready to conquer a new world. Little did I know that I would eventually form a life here like the one I always dreamed of, fulfilling my lifelong dream of becoming a wealthy individual. The next few years were not going to be easy, and the road to success was paved with many sacrifices, sweat, and tears. But it was *my* path and it made me the man who I am today.

THE BEGINNING OF THE ROAD

Once I arrived in the USA, I stopped in Jacksonville on my way to New York to visit a dear friend, Daniel. Driving away from the airport I saw something that determined my decision to stay in Jacksonville: the restaurant parking lots.

Daniel and I were driving down a busy thoroughfare in Jacksonville and I was surprised to see every restaurant parking lot was packed with cars and trucks, and it wasn't a weekend or a holiday. I questioned Daniel about this and when he told me it was always like that, I thought to myself, "There is a lot of disposable income here."

My conclusion came from a book I had read where the author theorized that an indicator of disposable income is the level of activity in restaurants, as people tend to avoid eating out during hard economic times. In time I'd learn the names of the establishments we drove by and realized how different they were in terms of cuisine, price, and quality.

From steakhouses to burger joints and everything in between, every restaurant and fast-food joint was packed with people regardless of the food they sold. I gained insight into the idea that there is a customer for every business and not every customer is the right one for you. It is a valuable lesson to learn so that we only spend resources on finding the right customers for our business. Many new and mature companies spend too much money advertising to the wrong target customer. Researching target customers and planning are vital aspects to a successful campaign *before* spending advertising dollars, thereby maximizing the reach of your advertising efforts. I would learn this the hard way … by spending lots of money on the wrong customers.

*****

My glamorous job offer at one of the largest tech companies in the world disappeared as quickly as my dreams of quick success in the USA.

The dot com bubble burst in early 2000 and took away my job offer. I was left stranded with no resources and no idea of how to proceed.

I knew the situation was my fault for making such a drastic move without any assurances, but now I had to face the music. The only thing that brought comfort was knowing there are no assurances of success in any enterprise you take; the only assurance you have is knowing you will never succeed unless you assume the risks and take advantage of the opportunities that come your way.

Regardless of this wisdom, I still had no money, no job, and no house. My family in Argentina needed to be taken care of and were waiting to join me in the USA.

Thankfully, Daniel took me in. I stayed in his home and he encouraged me to concentrate on mastering the English language first and dealing with my paperwork second.

It was the best advice he would ever give me as communication is the key to opening many doors. Without being able to communicate with others you can barely survive, you can't help others, and others can't help you as they don't know what you need.

I took Daniel's advice very seriously and in future years, applied the principle in my company. It was a source of strength within the business and allowed us to rise in the industry. Communication with customers is key to keeping them happy and to avoid misunderstandings.

The better a customer is informed on their project, the fewer chances there are of them becoming upset with the results. Remember, while you may understand your actions and the steps taken in a project, customers don't always understand them unless you communicate with them and keep them informed.

On July 14, 2000, I incorporated my first American company: Smarter Investments Corp. Later it became Smarter Remodeling and it was with this company that I transformed myself from being a young penniless Argentinian to, in a few short years, being in the top 10% of Floridian earners. A bit later, I entered the elusive 1% club. It took almost a decade to get there, and as much as I'm enjoying the final destination, the path traveled and the lessons learned along the way are the most valuable assets I have. Utilizing those assets is what allowed me to launch SR360 Solutions, a program to help other people and businesses achieve the same level of success I did.

Lesson of Success

— 28 —

Start.

Don't be afraid of taking the first step. Be afraid of the opportunities you will miss if you never decide to take that first step.

Lesson of Success

— 29 —

Regret and time.

I regret not taking more time to spend with friends and my extended family. Never make a choice that will cause regret. Make the time and do it now. Put your book down and call a friend, meet, go for coffee, lunch or a drink.

Why are you still reading? Put the book down and call a friend, *now*!

SCAN ME FOR MORE CONTENT

CHAPTER 11

BUILDING A REMODELING BUSINESS

I stumbled into the world of remodeling. I never intended to work in construction, but I was tempted by two partners who offered me partnership in a construction company. I was to run the administration side of the business while they would take care of all the construction side. Shortly after the partnership was agreed upon I found myself not only doing paperwork, but buying materials, assisting on the job sites, and talking to customers; all in all, a lot more than just paperwork!

A few months in, my partners decided to jump ship, claiming there was not enough money in the business for them to stay. They exited the deal, leaving me with a company in an unfamiliar industry. I had very little knowledge how to run and operate the business but had little choice as I had invested all my money into it.

I quickly learned the business's biggest shortcoming was we couldn't take larger jobs as we were not licensed. Luckily a person I met around that time, and with whom I would quickly become good friends, had a license and offered to qualify my company until I could get my own license.

Being properly licensed and insured opened many new business opportunities. One of the previous partners came back and began working for me. We quickly moved from the handyman business to large scale remodeling projects, but we couldn't make enough money to cover all the expenses and still have a decent profit. Once again, I found myself competing on price because we didn't have any other differentiator. I hired a marketing coach who taught us, "If you have no clear differentiators from your competitors, your customers can only choose based on price." Previous businesses had taught me that competing on price was a race to the bottom. So, I differentiated.

I immediately started making moves to create those differentiators. Customers wanted their projects done on time and were afraid of being taken advantage of by deceptive contractors; so, we began working on what became our *On Time On Budget Guarantee* ™.

We created a website showing customers our work and posting other customers' testimonials. This was one of the keys to our success and I can't stress enough how important this is. People trust the recommendations of people they have never met (aka online reviews) to make their purchase decisions.

There are many published books and blogs stressing the importance of word-of-mouth referrals. While I agree this is important, we live in a modern world where customers go to the Internet for information, so you should be very aware of what people can read online about your company.

A good word-of-mouth referral only reaches a few people, whereas an online review can reach thousands of potential customers. Work on building your reputation online as it will help you more than paid advertising and has a better rate of return per dollar spent. Read Chapter 14: Build Your Reputation Online for more tips on how to do this.

As our reputation grew and people could clearly see how we were different from our main competitors, they were willing to pay higher prices for better service and that was our turning point. We were finally able to stop basing our prices on what other companies charged and created a price structure that worked for us. Our newfound success provided profits to pay ourselves a salary and provided good opportunities for our employees and subcontractors.

Many authors with experience in the construction business say that you never base your markup on what other contractors are charging as you don't have their same overhead expenses. I agree with this. In the construction industry the concept that contractors should make 10% profits and 10% overhead is widespread. It's a notion that keeps being perpetuated by insurance companies.

Insurance companies put in their policies that they are willing to recognize up to 10% overhead and 10% profits for the contractors making the repairs on the properties they insure. In their infinite wisdom and generosity, they have determined all contractors can make a decent living with those numbers.

For the sake of clarity, allow me to repeat this once again: do *not* base your pricing on what others are charging. You need to come up with a number that makes sense to you and for your business.

The first step in pricing is determining overhead expenses. These expenses are any and all expenditures that cannot be assigned to a particular job, i.e.: rent, phone and internet expenses, insurance, office staff, vehicle expenses, marketing and advertising, equipment, tools, etc.

Once you know your overhead costs, determine how much profit you want. You will need to defend your prices with your customer, so ensure you have clear differentiating points from your competitors with strong benefits. Come up with a series of competitive advantages and sell your services based on value, not price.

We'll go in more detail about overhead expenses and profitable pricing in Chapter 13: Profitable Pricing.

At Smarter Remodeling we created On Time On Budget Guarantee™, Concierge Experience™, Skilled Trades Network ™ and others, which over time made us one of the top contractors in the state of Florida.

Lesson of Success

— 30 —

Sell on value not on price.

Selling on price rather than value ensures you will always be competing at the bottom. A business needs to be thriving rather than surviving.

Lesson of Success

— 31 —

Create differentiators.

Create differentiators between you and the competition. Customers have no other choice than to select based on price if you don't offer any differentiators. Every business is unique. Consider what you do differently than others and use that as a differentiator. It might be the services you

---

provide or the *way* you provide them. Whatever it is, make sure your customers know why you are different and what you do differently than your competitors.

Lesson of Success

— 32 —

Learn your margin, markup, and overhead expenses.

Learn the associated costs of your business and find profit numbers that work for you, and from there, set your prices.

Do not base pricing on what others are charging. This doesn't mean you don't need to remain competitive, only that you need your own numbers and if your price ends up being higher, then you either need to become more efficient (lowering overhead) or come up with clear differentiators that justify the price difference.

CHAPTER 12

———◆———

HOW TO START A CONSTRUCTION BUSINESS

Great! You have made the decision to start your own business. It's time to stop working for the boss man and start building your empire!

Now what? Well, there are a few boring things you need to do before you can begin your business, but they will prove critical to your success.

LEGAL ENTITY

Creating a legal entity may seem like a daunting task. It's best to get an accountant or lawyer to help with the details, but if you know what you need, you could do it yourself.

First up, decide if you want to form a limited liability corporation (LLC) or a corporation (Inc., Corp., etc.).

Each form has its own pros and cons, and a tax advisor is the best person to explain the difference. Once you select the type of company, see if you can elect to be a Subchapter S corporation to avoid paying taxes at the corporate level as well as at the individual level.

Once your company is formed, apply for a Federal Employer Identification Number (FEIN, also known as EIN). This is the tax number that will identify your company just as your social security number identifies you. This step can be completed online directly on the IRS website (for free!). There are multitudes of online services with domain names similar to the IRS website that will charge you to perform this task. There is no reason to pay somebody to do it unless you're uncomfortable doing it yourself. Regardless, you will need to provide identifying information for yourself and the business so the IRS can generate your EIN. The IRS generates this number immediately upon completion of the application. Save that number, as you will need it many times!

This is a very short overview explaining how to create a legal entity. Each company has their own set of documents that are important, such as Articles of Incorporation or Operating agreements. While for the vast majority of time you will never look at these documents, there may come a time where you have to; and if you don't have the right set, or have neglected to follow certain procedures, you might find yourself in difficult situations without the protection that a corporate shield allows you. Again, I highly suggest you consult an accountant or lawyer before creating your own legal entity.

BANKING

Once you have legally established your company and have an EIN, it's time to open a bank account. This step will be one of the easier steps in the process! When choosing a bank, look for an institution that offers services such as free checking accounts, debit cards, online banking, ACH payments, etc. Another important aspect is to look at the fees associated with the accounts as they will impact your profits.

It may be useful to choose a bank you have a preexisting relationship with, as they might be able to help you with financing needs you may currently have or will have in the future. Financing options could include:

Line of Credit

Having an open line of credit to cover cash flow shortages in your business is very useful. It eases the worry of finding money to cover expenses until your jobs make some progress and funds are received from the customer. Ideally it is best to carefully craft your payment schedules to prevent cash flow shortages. In the best scenario there will be no need for a line of credit, but it's better to have one and never use it than it is to need one and not have it.

Credit Cards

It's a very bad idea to use a credit card to finance your business operations. The interest rate associated with credit cards are normally much higher than a line of credit or a loan; however, if you are paying the bill in full every month, it may be a great tool to collect points or miles and use those to reward your employees, reduce the cost of acquisition of items you might need, or just to get some cash back. Look at the credit cards your bank is offering and decide which card fits your business operations best.

Merchant Services

If you are planning to accept credit cards (which is recommended if your average sale price is low) then ask your bank for the costs associated with this. You may be able to get a better price on the fees and the money from the credit/debit card transactions may transfer into your account more quickly than using a third party application. Tip: when calculating the cost of a job, include the cost of receiving payment via credit cards to the cost of the job to ensure you don't end up losing 2.5% to 4% (or more!) every time you receive a credit card as payment. Include this cost in all your jobs as you don't know how your customer will choose to pay and then offer a cash/check discount (or not) rather than telling your customer there will be a surcharge for using a credit card as a payment method (some credit card companies prohibit adding surcharges).

Loans

You may wish to inquire about loans to buy equipment or real estate. Unless you can provide good collateral, loans may be difficult to access if you are starting a business, but it is good to know what is available and what you need to qualify for one.

Most banks will require you to sign as a personal guarantor for your company. This means that if your company goes out of business you are still liable for the money the bank has lent you. Always try to get all your business debt secured only through the business assets and not your personal assets. Occasionally this may not be possible, at which time you must decide if you are willing to take that risk. If unsure, consult with a financial or tax advisor before pledging personal assets as collateral for business debts.

As a side note, find a bank where you can build a relationship with your personal banker, as this may be of great benefit in the future.

INSURANCE

Once you have a business account and have deposited some money into it, it is now time to obtain a few different types of business insurance. This could be very simple or extremely complex, depending on who you ask. If you have a trusted commercial insurance agent, ask the agent for quotes for the following policies: general liability, workers' compensation, and, if you have any company vehicles, auto.

General Liability

General liability insurance provides coverage for certain liabilities which might incur during the normal course of business. An example is accidental damages to a customer's property.

Determine the extent of your coverage. Nowadays, most contractors use a $1,000,000 per occurrence limit with a $2,000,000 general aggregate. Lower limits are available, but the price difference is not significant and it is wise to be protected as much as you can afford without it causing financial hardship. Choose your plan carefully, as it is unnecessary to pay for a $5,000,000 policy if $1,000,000 will cover the vast majority of your work.

With the general liability policy you can obtain additional riders. According to Investopedia.com, a rider is an **insurance policy provision that adds benefits to or amends the terms of a basic insurance policy.** Riders provide insured parties with additional coverage options, or they may restrict or limit coverage to the additional coverages you may need. This policy covers you but is also a protection for your customers as it provides them with peace of mind. They can rely on an insurance company to pay for damages you might cause, rather than attempting to get that money from you.

Workers' Compensation Insurance

This insurance covers you when an employee gets injured on the job. Some states allow exemptions from this insurance if you have no employees. This exempts the owners of the company from getting workers' compensation insurance and if they get injured, they won't have coverage. Most small businesses without employees decide to be exempt to save a few dollars. They are betting they will not get injured, or if they do, it's a small mishap. It's relatively easy to obtain the exemption for owners, but remember, this doesn't exempt employees.

The cost of a workers' compensation policy depends on how much payroll you have and what risks are associated with your business. For example, a policy for roofers will be more expensive than a policy for interior carpenters. If and when you use the policy, your expenses will increase significantly. Always include this number when calculating your overhead and/or costs.

Auto Insurance

Auto insurance covers the liability and repairs to damaged company vehicles. It will also pay for damages your vehicles cause to other people and their property. Extra coverage can be obtained to cover your vehicles, rented vehicles, etc.

*****

A general rule of thumb with insurance policies is to get as much coverage as you can afford without it negatively impacting your bottom line. This will afford you more protection and coverage when something goes wrong. The internet is plagued with stories of people who thought they were covered but their insurance limits were too low and they were held responsible for the difference!

There are other insurance policies you may need, such as property insurance if you own a building, or flood insurance if your building is in an area prone to flooding. Another good value policy is an umbrella policy. An umbrella policy adds additional limits to existing policies and is often more affordable than increasing the limits on existing policies. For example, if you have a $1MM ($1,000,000) limit in your general liability, vehicle and workers' compensation policies, then if an umbrella policy is added that covers all three of those policies, you can increase your limits to $2MM ($2,000,000) on each policy. Ask your insurance agent for more information.

This is not intended as a comprehensive list of all the insurance policies you may need, but rather is the minimum coverage you might need to get started.

I'm also not advocating wasting money on unnecessary coverage, but be aware of the limits offered by your insurance agent/company and research how much more it will cost to increase coverage. If it's not too much and you can afford the increase, then do it.

BUSINESS LICENSES

To operate your business, you will need a business license, which is a tax paid to local municipalities. A local registry office or municipal office can help you with the business license(s) you need. There are some jurisdictions that don't require you to have a business license. Some other municipalities will require you to have a personal business license as the license holder in addition to your business having one. Check with the local authorities to see if you are required to have one.

PROFESSIONAL LICENSES

Professionals are often asked for their credentials. In the construction world, common professional licenses include the general contractor license, building contractor license and the residential contractor license. These are the denominations used in Florida. In other states the equivalent licenses might be called by different names.

Your professional licenses are more difficult to obtain and require preparation (studying!) before you can take an exam. Every state has their own requirements and you must meet their individual requirements. For most trades such as carpenters, drywallers, and flooring installers, the requirements to obtain a license (if needed) are simple and straightforward. For electricians, plumbers, or AC contractors, the requirements are usually in line with a general contractors' license.

*****

Always make sure you have the proper licensing to perform your work, because without it you could get fined or shut down. Most states don't offer any protection against homeowners who don't pay their bills if you performed work that needed a license but you operated without (unlicensed activity). In fact, in Florida, a homeowner could get away with not paying what he or she owes you if you have performed work on their property without holding the proper licenses. It is not worth the risk. Always perform work following all the current state and local building codes and comply with local ordinances.

Requirements for each state vary greatly so we have only broadly covered the necessary licensing: *Do not perform unlicensed work.* If you need a license, take the time and get it. It will pay for itself many times over.

Lesson of Success

— 33 —

Find a personal banker.

I cannot stress how important this is. I have been lucky to be able to establish a relationship with an excellent personal banker. Having an established relationship removes numerous financial stresses. Your personal banker is your personal champion inside the institution. He or she knows you and your business, so they can use financial records to paint a holistic picture of your business rather than a cold spreadsheet. Personal bankers advocate for you and your businesses' best interests in front of the bank management and underwriters. This person will be fundamental to your success. Find this person soon.

Lesson of Success

— 34 —

Find an insurance agent.

It will greatly benefit you to find an insurance agent you can trust and who can offer sage advice. It is important to find an agent who has extensive experience in the construction industry and who is familiar with all the clauses and exclusions that may cause you trouble down the road.

Agents work on fees they get from the insurance companies and the more policies you have with one agent, the more you mean to his income, and the more attention he can devote to you. Shop around for the best prices but always allow your agent to have the last chance to price a policy for you.

Establishing a relationship that's beneficial for you and your agent will provide you incredible returns. Your agent has the contacts inside the insurance companies that can really help you with your policies or any issues you might have with them.

Scan the code to get access to some of my contacts.

SCAN ME FOR CONTACTS

CHAPTER 13

PROFITABLE PRICING

Let's make money! With all the legal requirements taken care of, we can start talking about how to make money so you can have some fun.

I'm referring to the money you keep for yourself, which is also known as *profit*.

Profit differs greatly from revenue. Revenue is the money that comes into your business, but not all of that money is yours. You still have bills and employees to pay. However, profit is *mostly* your money … Uncle Sam wants his share too! The good news about paying taxes is that you are contributing to paying down the ever-expanding bills of the United States of America, but it's more of a benefit to yourself, because owing income taxes means you are making money and that's always good.

So, how do we get to make money? First decide what your markup will be. Markup is the number you will use to determine what price you need to sell your goods or services for. In order to know that number we first need to know a few key pieces of information.

ANNUAL REVENUES

We are not fortune tellers and while you don't know how much your annual revenues will be, you still need to come up with a good estimate. A good estimate begins by calculating the number of goods or services sold throughout the year; that will suffice for now. You may be looking at your book or your speakers and thinking, *This guy is crazy! How would I know that? I just wasted the money I paid for this book!* (By the way, I hope you did pay for the book and your copy isn't one of the one hundred my mom bought!) But bear with me. It is difficult to create an estimate for what your revenues will be, especially if you are new in business. First, begin by determining what volume of work can you realistically complete every week, taking into consideration whatever else you do for the company. For example, let's say you renovate bathrooms and are a one-man operation (God bless your soul). Take the average amount of time it takes to renovate a bathroom, minus the time you spend collecting materials and chasing sales. How many bathrooms can you complete in a week, on average? While no two bathrooms are the same (or so you'd think), for this example we say half a bathroom per week. This comes to approximately 26 bathrooms per year.

Second, determine the cost per bathroom. For the sake of our example, we will say that the average sales price of a bathroom remodel is $15,000.

Third, use the volume of workload and multiply that by the sales price. Completing 26 bathrooms per year at a sales price of $15,000 each equals $390,000.

If this is your only source of revenues, then $390,000 will be your annual revenues. If you do have other sources of revenues, then add them to your bathrooms' revenues and that will be your annual revenues number.

You could also estimate your numbers or get them from conversations with other contractors, but good luck trying to get other contractors in your market to cough up that information!

To summarize:

Annual Revenues = (Volume per week) x Price x 52

OVERHEAD EXPENSES

Your overhead expenses will be all other expenses you have that cannot be attributed directly to a job. For example:

- office expenses
- in-office personnel salaries
- utilities
- telephones
- vehicle expenses
- insurance
- accounting and legal fees
- janitorial supplies
- licenses
- tools

A side note on tools: tools that are used in many jobs should be considered as an overhead expense. As a rule of thumb, Smarter Remodeling calculates all our tools purchased as overhead and all the consumables (blades, discs, etc.) as a job cost.

The owner's compensation (more about this in Chapter 32) and salary will also be an overhead expense with the exception that if the Owner's salary comes from work he performs at the job site, then this will be attributed to the job cost as Labor. Otherwise, if the salary is for administrative tasks performed (meaning it cannot be attributed to just one job), then this salary is an overhead.

Many overhead expenses will be paid monthly (vehicles, rent, phone bills, utilities, insurance) and some may be paid weekly (salaries). To get a relatively accurate number, multiply all the weekly expenses by 52 and then divide that number by twelve. Your answer will be your monthly overhead expenses. Add that to your other monthly expenses and you have your total monthly expenses number. For our example we'll use $10,000 per month.

What I like to do is convert that number into a daily number, as it is a much simpler unit of measurement. To arrive at a daily overhead cost, take the monthly overhead cost of overhead and multiply it by twelve ($10,000 x 12) and the result is $120,000. Now we divide that number by the number of working days in the year. The average number of working days in a year is 250 (365 days minus weekends and holidays).

Following our example then, $120,000 gets divided in 250 working days and we get $480 per day as our overhead cost.

Determining overhead expenses is vital to calculating overall expenses and will help you to create a more accurate financial picture. It is also the basis to establish an annual budget that will keep you profitable and on track. Spend whatever time you need to get accurate figures, because it will be worth it.

To summarize:

Daily Overhead Expenses = ((Monthly expenses x 12) + (Weekly Expenses x 52)) / 250

COST OF GOODS (OR SERVICES) SOLD

This is a very common account you will find in your accounting program's Chart of Accounts. We will simplify it here and call it *Job Costs* to make it more understandable and relatable. Here you will calculate all the costs directly attributable to a job. Information needed to calculate your job cost are:

Labor

Labor includes the costs of both employees and yourself (if you don't count your salary in the overhead expenses). Ensure to include all the expenses associated with an employee that are not included in the overhead such as payroll taxes and benefits (vacation, time off, sick time, and 401K employer contributions). Note: If you run your own payroll, you probably have accounts to track these, and if you use a payroll service they most likely will include all the expenses in the amount you pay every pay period.

Materials

This includes everything you will use to perform the job. Common materials will be lumber, drywall, tile backer, screws, nails, glue, paint, and tiles. Anything physical you will use in this job must be included here. At Smarter Remodeling we normally buy large boxes of screws or nails as they are more cost effective and although we often don't entirely use them all in one job we attribute the cost to a particular job. Technically we should determine the cost of what we use for the job but that becomes too complicated and is not an efficient use of anyone's time.

Subcontractors

This cost includes any amount paid to subcontractors (plumbers, electricians)—anyone who's working on the job but is not an employee needs to be accounted for here.

Others

Many jobs have unique expenses or irregular expenses. This would include things such as permit fees, dumpsters, cleaning supplies, consumables, equipment rental fees, and delivery fees.

To determine the job cost, add everything up. For example,

Labor: $2,215 (For the purpose of our example we will say this covers four days of labor)

Materials: $1,100

Subcontractors: $800

Others: $575

Total Job Cost = $4,690

To summarize:

Total Job Costs = Labor + Materials + Subcontractors + Others

BREAK EVEN POINT OR REAL JOB COST

How do you know what you need to charge to cover all your costs (job costs plus your overhead)? Once you know this number, you will know the breakeven point for the job. Some people also call this *Real Job Cost,* as this number contains everything you need to have an accurate and all-inclusive cost.

Let's use the numbers from the examples above to determine our breakeven point. The number of days it will take to complete the project is four.

Job Overhead = $480/day x 4 days = $1920

Job Costs = $ 4,690

Job break event point = $1920 + $4,690

Total cost = $6,610

Now you know that to break even you need to charge at least $6,610. Since we are not in business to break even, we need to add some profit margin to make it worth our time.

To summarize:

Break Even Point = Job Overhead + Job Costs

PROFIT

Profit is the amount of money you want to make on top of your break-even cost. Since you already know what the real job cost is—which included your overhead expenses (this also included your owner's compensation for your risk in the company and your owner's salary for the work you do for the company), then the profit is the amount you want to keep when a job is complete and the expenses are paid for.

Let's say you are a very reasonable person and only want to make a 10% profit. (Remember this is after paying *all* your expenses.) Following the example we've been using, you add 10% to $6,610.00 and you are done.

Well … not exactly.

If you want to keep 10% then your markup (the number you will use to get there) needs to be larger than 10%.

10% of $6,610 is $661. So, if you add 10% to $6,610 then the final price will be $7,271.

Now you see the final price (or sales price or retail price).

Suppose a few days later you decide to spend your profit. Since your profit rate is 10% you determine $727.10 and some change is your spending money.

Hold on a moment, cowboy. You've created two numbers that 'equal' 10%: $661 and $727.10.

Deduct that $727.10 from $7,271 and you get $6,543.90 or $66.10 less than your break-even cost. What happened? What kind of magic is this?

No worries. There is no magic here, just simple math. When using percentages with different base numbers, the results cannot be the same.

Use the following formula to determine the margin percentage:

((Sales Price - Break Even Cost) / Break Even Cost x 100)

Using our example, we can calculate the margin percentage we obtained:

(($7271 - $6610) / $6610) x 100

(661 / $7271) x 100

0.09090909 x 100

9.0909

As demonstrated above, the margin percentage you get from using a Markup of 10% is only 9.09%; a figure slightly less than 10%, if you want to take 10% from the top, you need to increase your Markup percentage, in this case to 11%.

For clarification purposes, Markup is the percentage you add to your costs to come up with a sales price and Margin is the percentage from the sales price you get to keep. In this case your profit margin will be 9.09%. This will give you a slightly lower amount ($660.93) than the profit amount you added to the breakeven cost ($661) but it's only because we rounded the decimals.

If you want to put 10% in your pocket from your total sales, you need to use a larger markup than just 11%. Why? You still need to pay taxes on your profit and other expenses (which always come unexpectedly during the year for things you didn't account for). Hopefully you can control unexpected expenses to some degree, keeping them to a minimum but they will be some. Unexpected expenses, or 'unbudgeted expenses' as many people called them, are a fact of life in every business and especially in the construction business.

Use the formula shown to calculate the profit margin percentage you are getting from your sales. Aiming for double digits after accounting for taxes and unbudgeted expenses should be the goal; I would recommend you begin with a minimum of 20%. Once again, this is above your breakeven cost which includes your overhead expenses.

As your business grows, becomes stable and older, you will learn to predict certain variable expenses, such as tools and repairs to vehicles or buildings. Your overhead expenses will become easier to define and consistent, providing you numbers you can be more confident in.

This method of establishing your price assumes you know how long a job will take but it doesn't account for times when jobs take longer than expected. Many contractors don't worry about this as long as the delay doesn't increase the job cost. Normally it doesn't but it does increase the break even and real job cost.

Let's take a look at the numbers again to see how a delayed project negatively impacts our bottom line.

Let's go back to our example. You have a small job that you predicted and priced using four days of labor. The job cost was $4,690 and the overhead expenses $1,920 for a break even cost or real job cost of $6,610.

You started the job on Wednesday and on Thursday, you receive a call from the homeowner stating they had a death in the family and won't be home on Friday but will be back on Monday. You express your condolences and tell him you will be rescheduling your guys to go back to the jobsite on Monday. On Sunday they call again stating there have been delays and will be back in town late Tuesday evening and your crew can restart on Wednesday. You once again express your condolences, thank them for the notice, and reschedule your guys to Wednesday at which point the job restarts and completes within the anticipated four days.

Your job cost didn't change. Your labor, materials, and other costs were the same. But what could have changed is your overhead expenses not the actual amount of the overhead but the number of working days you use to calculate your daily overhead cost.

If you were able to put your crews on a different jobsite on Friday, Monday, and Tuesday then nothing changes and all the numbers remain the same. However, if your crew had to stay home those three days and you had no other revenue coming in on those days that would pay for the overhead expenses, then you now have three days less in your number of working days for the year.

We used $120,000 as the annual overhead expense and 250 days as the working days for the year. When the annual overhead expenses were calculated by the number of working days in the year, the overhead cost per day was $480. But since three working days were lost, the number of working days in the year was reduced to 247 which drove up the daily overhead costs.

Does it matter? Probably not. It's just a few dollars, but if this happens 20% of the time then those few dollars will start adding up quickly. One easy way to compensate for this and to somewhat account for a number of days you won't be working (for whatever reason) is to deduct a certain number of days from the working days to compensate for this.

My suggestion is to deduct 50 days from the number of working days and use an even 200 days. This will allow you to "lose" one day a week, and since most weeks you won't lose a day, it will accumulate for the weeks when more than one is lost. With this new number of working days (200), divide $120,000 by 200 and your new daily overhead cost is $600.

Let's recalculate the new break-even cost.

Job Cost = $4,690

Overhead Expenses = $2,400 (The new rate of $600 x four days of work)

Break even cost = $7,090

Markup percentage = 11.00%

Final price = $7,869.90

The price difference between your first calculation and the second is only $598.90, or approximately 7.5 % of the final price.

Your business will be better served by using the reduced number of working days for the year as this allows overhead expenses to be covered in a more reasonable amount of time. It also does not impact your prices significantly.

You can use the following table as a reference so you know what markup to apply depending on the profit margin you want to achieve.

Markup %	Profit Margin %	Markup %	Profit Margin %
5.00	4.76	55.00	35.48
10.00	9.09	60.00	37.50
15.00	13.04	65.00	39.39
20.00	16.67	70.00	41.18
25.00	20.00	75.00	42.86
30.00	23.07	80.00	44.44
35.00	25.93	85.00	45.95

40.00	28.57	90.00	47.37
45.00	31.03	95.00	48.72
50.00	33.33	100.00	50.00

RESERVE ACCOUNT

It's very good idea to establish a reserve account that gets automatically funded with every job. Begin by putting in 3 to 5% of every job to build it up quickly.

Once this reserve account reaches six months of overhead expenses, then drop the contribution to 1 to 3%. Finally, when the account reaches an amount equal to or larger than one year of overhead expenses, drop the contribution to 0.5 to 1%.

The purpose of this account is to provide security and act as a lifeline when cash flow is slow or non-existent. Instead of borrowing money you can use the funds in the reserve account to cover operating costs. Even the best of planners need additional money that's not in the business account (and when that time comes you can send me a thank you note for encouraging you to establish this account!).

Ideally, the reserve account should be funded immediately as deposits are received into your main account. Automatic transfers are fantastic and are the easiest way to complete this task. If automation is not available, create a routine of doing it manually the same day a deposit has been credited to your account. Don't put the task off and don't use this account for anything else than to deposit reserve money.

Lesson of Success
— 35 —

Know your numbers.

Understanding the financial numbers and statistics of your business is *critical* to your success!

If you don't understand these basic business formulas and numbers, and if you can't keep them under control, you will be out of business within a short amount of time

If you understand the numbers **or** if you are comfortable controlling them, you will be out of business in a short amount of time plus a day or two.

You must understand **and** control your numbers, and only then you have a chance to stay in business for many years!

Know your numbers!

Lesson of Success
— 36 —

Establish the right price.

Establishing the right price is critical to success, but don't make the mistake of thinking higher prices are the solution.

For prices to be enticing to consumers, they must provide lots of value. Your prices need to be competitive (but that doesn't mean lower than competitors). More importantly, they need to provide value your customer sees and understands—if your customers cannot perceive the value they will only judge you based on price.

Lesson of Success

— 37 —

Explain price increases.

If your profit margins are declining, your first instinct will be to raise prices. This might be acceptable in market conditions like increased cost of materials. Be sure to explain to your customers why you are raising prices and then create added value for them.

For example, in the early part of 2020 materials costs increased by 20%, but at Smarter Remodeling we only adjusted our prices by 7%. Customers often don't realize your prices include much more than only materials but they can see value by not being charged an additional 20%.

Lesson of Success

— 38 —

Understand company efficiency and overhead expenses.

If profit margins are declining, look for areas to improve efficiency in your company. Ask yourself, do I have too much overhead for the revenues I'm generating?

Lowering your overhead will allow you to increase your profit margins without raising prices.

CHAPTER 14

———◆———

BUILD AN ONLINE REPUTATION

Building your online reputation may look like a daunting task but it's much simpler than it looks. A key factor in building a lasting online reputation is consistency and not falling behind on tasks needed to maintain your reputation.

By far, the most viewed online reviews will be the ones your customers will post on Google Reviews. While the space may change over time, as of today, Google Reviews is the dominant player in this arena.

Start by claiming your Google Business listing and completing your profile. Add plenty of photographs and videos of completed jobs and continue adding content as your portfolio expands. Keep the listing updated with business contact details, hours of operation, and a working link to your website.

Second, use social media such as Facebook business pages, as many customers and potential customers will take the time to look for you on the various platforms. You can repeat the content posted on your Google business listing, but remember that content is king. You need to be prepared to generate good content for your social media platform and ensure this content is relevant and interesting to your target audience.

Finally, prepare step-by-step instructions for your customers to follow from a computer and from a mobile device (as they follow slightly different processes, create both methods) so they leave you a review. Have these instructions ready to send by email but also have some printed instructions available.

Upon the successful completion of a project, make sure your customer is satisfied and establish a conversation about the project and how it went. Speak about the importance of customer feedback for your company and ask questions aimed at eliciting specific responses.

Here is an example of what could be said:

Business: Are you happy with how we handled your initial contact with our company?

Customer: Yes! The lady who picked-up the phone was extremely friendly and answered all my questions. She was even able to schedule a consultation the next day.

Business: Great! And how was your consultant?

Customer: Paul was great; he had many good ideas for the project and provided a very detailed proposal that made us feel comfortable with the expense involved.

Business: That's good to hear. How was your experience with your project manager?

Customer: Very good! He kept me informed of everything that was happening at all times and made sure the job site was always tidy and clean.

Business: That's fantastic to hear of. Would you say the project was completed to your satisfaction?

Customer: Absolutely!

Business: Are you happy with your project and the way our company handled it?

Customer: Of course! I'm very happy! You guys were fabulous!

Business: So, it's safe to assume you would refer us to your friends and family?

Customer: Most certainly. I can't wait to show off my project.

Business: So, let me resume to see if I got everything right. What I heard you say was that from the initial call through to the consultation phase [INSERT COMPANY NAME] was very accommodating and provided a professional service that made me feel comfortable to hire them. Once the project started my project manager kept you updated on the progress and any issues that came up. As well, he kept the jobsite clean and completed the project to your complete satisfaction. You were very happy with the experience and would recommend [INSERT COMPANY NAME] to your friends and family. Is this an accurate summary of what you just told me?

Customer: You understood correctly. That is exactly what I told you.

Business: As a locally owned company we depend heavily on the referrals by our satisfied customers to obtain new projects. This allows us to save on advertising and in return we pass these savings to our customers. I just texted/emailed you a summary of what you just told me. Would you allow me to show you how to post that as a review on our Google/Facebook/Yelp review page? The only thing you need to do is copy and paste it onto the page.

At this point, you should have a draft email prepared with all the links to the review pages you want the customer to post to. Using the information received from the customer, create the review and send it to the customer. By asking the preliminary questions and summarizing the review for them (and using their own words), you just increased your chances of getting a positive review by 95%.

Most people don't post reviews because they don't know what to write, or where, or how. Also, most customers won't post a review if they don't do so immediately.

Use the above questions as an example of what could be said and done. Adapt the questions to your particular company so you get the answers you need. Having a process in place like the one demonstrated above will increase your reviews tenfold.

Besides Google and Facebook, find other local review sites for contractors in your local city, county, and region.

A recent player in the neighborhood space is Nextdoor. This app creates community spaces revolving around individual neighborhoods (or communities) and allows neighbors to interact with each other, make recommendations, and post news. It has become very popular in the last few years as it creates a space for locals to share. If your customer is familiar with Nextdoor, ask them to post the review there as well. There are opportunities to find potential customers with similar profiles to the one posting the review so you may get some additional business from using the app. Other less popular sites that are still relevant are Yelp and Angi (former Angie's List). However, the majority of your online customers will find you through Google and Facebook.

When asking for reviews from your customers, ask if they use the local Nextdoor app and if they would post their review for you. If they don't, then ask if they would post it to your business's Google account and give them the link to it. Always provide a link as it is easier for your customers to click and start writing, which will result in more reviews.

Do not assume your customer knows how to post reviews and that they will do it in the future. Have a system in place for them to do it immediately after completion of the job.

An example of a great review system for collecting reviews was in my eye surgeon's office—he is one of the top eye surgeons in the world and he makes sure everyone knows it. Once the surgery is completed, they take the patient into a recovery room and the doctor comes in to do a checkup. For most patients the results are immediate; and the client is in a state of amazement at the genius who gave them back their vision. Once the checkup is completed, an assistant takes the client to a second room where they sit down with the client and ask for the review. If the client tries to do it later, they state that as the client has a few more minutes before they can sign the discharge papers, so now is the perfect time to do the review.

They ask you to open your Facebook or Google account but as the client's eyes are still covered, they "volunteer" to do this. And they describe the process using fancy and sophisticated terms making it sound like poetry and positioning the eye surgeon as a "vision god" who valiantly fought for your vision and slayed a dragon on your behalf. They read it back to the client, who is thinking "Wow, I'm a good writer!" The client nods for their approval and it's done. They got another stellar review.

While I'm not a fan of this "semi-forced" process, I am a big fan of providing an already written message using the customer's words as it simplifies the task for your customer, and it works! Your customer doesn't need to think about what to write: you did it for them using their own words.

Over the years I have used several methods to collect reviews and this method has been the most effective

It never hurts to ask for a review, so ask for the review to be posted right there. If they decline, don't be pushy but follow up and when you do, don't just ask for the review. When following up with a client, check if everything is still fine with the project and ask if there is anything else you can do for them adding a note at the end to the effect of, "By the way, I know you are busy but could you please take a moment to post the online review? Reviews mean a lot to us and we would greatly appreciate it if you would take a moment of your time and do so now. Here are the links for the different sites so you don't have to waste time looking them up. Below the links you can see some of our other customers' feedback if you need inspiration."

Often customers that haven't placed the review will do so when you follow up with them. Of course, there will be some who won't, so you can remind them in your next follow up call.

Typically, construction businesses should follow up with customers after completion at 30 days, 90 days, six months and one year. These calls are not specific to ask for a review but to check on the project and to know if everything is still satisfactory. If the customer hasn't placed a review, you can use the opportunity to ask or remind them. These calls are also great opportunities to ask for a referral.

Always monitor the sites where you ask your customers to place reviews and make a habit of responding to them.

Even bad reviews can be turned into positive reviews if you respond to them properly.

Don't engage in an online war with unsatisfied customers. Many business owners believe they have to show these customers who's boss and put the client in their place. Never do that. It will only put off people reading these reviews. Instead, offer a calm, reasonable response promising to look into the complaint and ensure your company is doing everything it has promised to do.

Some customers try to use online reviews to extort free goods and services from you. They know how important online reviews are and they are willing to take advantage of that fact. Don't let them. Reply to the bad reviews without engaging in a war of words explaining your company's position without attacking the customer. This shows other customers you handle issues in a professional manner.

Focus on reviews *more* than advertising, as they are (relatively) free and very highly regarded by potential customers. Reviews are not 100% free if you have to assign an employee to look at them daily, but, it is very much like the old-money saying, "It may not be the key to happiness, but it is way ahead of whatever is in second place!"

Lesson of Success

— 39 —

Use technology to develop an online presence.

No matter how much you may or may not like technology and social media, the world has evolved and it's time to embrace these changes for your business.

If you don't want to run this aspect of your business, then consider outsourcing it to somebody you trust, but it is very important that you develop an online presence strategy. After all, like it or not, your potential customers are looking for you there.

SCAN ME FOR MORE CONTENT

CHAPTER 15

———————◆———————

BUSINESS OR JOB?

Now that your company is established, the insurance is in order, a shiny business credit card is in your wallet, the markup numbers have been determined so you can make a profit, and jobs are rolling in, it's time to assess if you have a business or just a job.

Let's start by defining how a business is different from a job.

When most people start their own business, they do all the tasks: getting leads, answering the phones, estimating, material sourcing, delivery, permitting, job execution, and payment collection. You are a one man show and do it all. The phrase "the buck stops here" was coined for you.

But doing all the work doesn't bother you. You are a proud business owner. You are at the top of the food chain now. There are no more bosses to answer to, you set your own hours, and make as much money as you desire. Sound familiar? I hope you are one of the extremely go-getter individuals who answered yes to all those statements. But if you belong to the (regrettably) extremely large group of business owners who cannot do that, then keep reading.

The best method to determine if you have a business or just a job is actually quite simple: ask yourself the question, if I was suddenly kidnapped by aliens, would my business still be operating without me? In other words, would things still get done? The alien scenario may be a bit of a stretch, but replace it with whichever analogy better fits your narrative (illness, travel, emergency etc.).

If you cannot attend to any of the tasks that makes your business operate, would the business continue moving forward? If the answer is yes, congratulations, you have a business! If the answer is no, pay attention to the following paragraphs to learn how to transition from a job to a business.

This definition is not 100% clear cut. A business may continue working without you being physically present at the job site if you do remote tasks such as bookkeeping, payroll, and billing. In this scenario, you have other people doing the heavy lifting for you (either employees or subcontractors). This is good, as it is a stepping stone to having a fully independent business that you own, but that is operated by others.

In other words, you should control your business but it doesn't require your skills to operate. Are you feeling a bit left out, thinking, why would I want to be made irrelevant by my own company? You may even be thinking, this guy smelled too many garlic bulbs when he was a kid and now he's just insane!

Defer your judgment for a few moments. I'm not disputing the fact that I might be insane, but give me a chance to convince why you will want your business to operate without you.

If your goal of owning a business is to have the resources to spend more time doing the things you enjoy doing, such as fishing, traveling, attending sporting events, spending time with your spouse, volunteering, etc., then you want a business that produces profits without you, allowing you to use the profits to keep doing the things you want to do. To do this, you need a team of people who operate the business, producing revenues and profits. If you don't have a full-functioning team, refer to Chapter 17 on creating an Organization Chart to learn how to create and develop an efficient team.

Lesson of Success

— 40 —

Have a business, not a job.

Your aim should always be to have a business not just a job. It's fine to have a job inside your business but it's important not to be consumed by it.

Early on, when establishing your business, work to create systems and procedures and hire the right talent so your business can run without you.

This is not an easy process, but it's completely doable, and has been done by millions of business owners who have already done it.

Remember, get started. It's one small step at a time.

CHAPTER 16

CASH FLOW—ACCESS TO CAPITAL

One of the most difficult things Smarter Remodeling had to overcome was the lack of access to capital. Some people think this is not important, but often it can make or break a business. Many people confuse access to capital with loans, and while this is one way to access capital, you can also achieve it by creating a proper schedule of payments from your jobs that allow for a positive cash flow by using your customer's money.

SCHEDULE OF PAYMENTS

Let's delve deeper into this. Many contractors operate by taking large deposits from customers on *future* jobs that allow them to complete *current* jobs (and thereby get paid for it), while in return they use that money to start the job from which they initially got the large deposit. The problem is that this is not a sustainable business model. It only takes one missed payment for your business to go down in flames. It's hard to resist the temptation of using deposit money to cover other expenses, but that money is not yours … yet. You can stop looking at your book or your speakers thinking I have lost my marbles. I haven't … yet!

Let me explain. A customer signs a contract with you to begin a job on their property and gives you a deposit so you can buy materials, engage subcontractors, start architectural drawings, or any other activity that needs to be done on *their property*. Even when the contract may grant you legal rights to the money, those funds should only be used for the job they were intended for. Misusing the money is asking for trouble.

I've seen really good contractors go down because they mismanaged their money or spent it on things they shouldn't have.

I remember a contractor I knew was offended when I called out his brand-new pickup truck (a few weeks earlier he had been complaining about not having enough money to cover his expenses as one of his large jobs was delayed). The contractor told me he had a right to drive the new truck.

I replied, "I'm not implying you don't have the right, only that I'm surprised you can afford it when just a few weeks ago you were complaining about your money woes."

"That's just the past," he said. "I just sold a new large job and got a big deposit."

I was floored. "Please don't tell me you used the new job to cover the expenses of an old job, and on top of that, you also used it also to get a new truck!?"

He got upset that I was questioning his decisions.

"Well, it's my money," he told me.

"No, it's not your money, *yet*. You haven't earned it *yet*, as you haven't performed any part of the new job. You should consider that money as belonging to your customer until the moment you have done the work to have earned it. Only then can you confidently say the money was yours."

The contractor didn't agree with me and said that as soon as his previous job paid him, he would get the new one going. Regrettably, his other job never paid him and the customer for the new job ended up suing him for the funds which he had already spent. Shortly after, he went out of business and lost everything, including the brand-new truck.

The moral of the story: *do not* use deposits from one job to pay for unrelated expenses. Hold deposits in a separate account until you are ready to start on the job and start earning that money. When you do this, combined with a good schedule of payments, you'll seldom have cash flow problems.

On the other hand, if you allow the customer to dictate the payment terms you are guaranteed to run into cash flow problems.

One time, after spending over a week going back and forth with a customer over a proposal to build an addition, and after we had agreed on the price and the scope of work, I submitted the contract for a signature along with the schedule of payments requesting a 20% deposit plus additional milestones payments as follow:

- 20% concrete
- 20% framing
- 20% roofing
- 10% rough electrical and mechanical
- 7% flooring
- 2% punch out list, and,
- 1% final payment (more on this later).

The contract at the time was pretty extensive for Smarter Remodeling, coming in at about eight pages (today they are over fifteen) so I wasn't surprised to see the customer was taking some time to read it. But I was flabbergasted when he told me, "I'm not going to sign this contract. There is no way I will finance your business."

Reeling from the surprise, I asked, "Finance my business? How's that?"

He defiantly pointed at the contract and replied, "You want me to give you a 20% deposit and then also to pay for framing, concrete, etc."

My incredulity to his complaint must have transpired over my face because he added, "You honestly believe I will pull money out of my pocket so you can build this addition without using your money?"

At that point I wasn't sure if it was OK to laugh or if it was better to walk away.

"Sir," I said. "I understand your concern but what I don't understand is why you believe I'll use *my* money to finance *your* project. Furthermore, the deposit is congruent with the specialty items you want for your project which cannot be returned for any reason; thus, you should understand I need to cover those costs plus the additional expenses of getting the drawings drafted, permits, and mobilization expenses. The rest of the money is being requested as the job reaches certain milestones as specified in the schedule of payments. In other words, we cannot move further without this deposit and your agreement to the schedule of payments. In actuality, we shouldn't move forward with the project at all since it seems we have widely differing opinions on how this project should move forward. You are better off hiring another contractor that's willing to work on your terms."

With my speech complete I started gathering my things and after expressing my gratitude for the opportunity I started walking away from the job.

As I was doing so, he told me, "You'll never make it in this business if you walk away from jobs like this and much less so when you try to get customers to finance your business!"

I looked at him with a smile on my face and finished the exchange with, "Thank you for your words of wisdom. I'll make sure to remember them."

As you can see, I didn't lie to him. I still remember those words, and doing exactly the opposite of his advice has made me a successful contractor. I never use my own money to finance customer's jobs, and I never let customers get behind with the schedule of payments. In other words, I always collect more than what I have performed so far and I always set aside the money I still haven't earned. This has allowed me to always be able to pay my bills on time to my vendors and subcontractors, which has been key to developing a solid network of tradespeople.

FINAL PAYMENT

On the subject of the schedule of payments, never let the customer dictate the terms of the agreement. In other words, you have to determine when you need the money from the job and get the customer to agree to those terms. If they propose terms that are not favorable to you, walk away. Do not finance customers' jobs with your own money. *Never.* If you do, you'll regret it sooner rather than later. You can offer financing through a third party but, once again, don't use your own money to finance a job. If you use a third party to offer financing, ensure they are releasing the money in the way you need it. Many financing companies only release the money when the job is completed or require the homeowners to sign off before they release your payment. Allowing the customers the power to hold your payment is a recipe for trouble.

Many contractors leave 10% of the job's payment to be collected at the end (final payment). In my opinion, this is way too much money to leave to the whims of your customers. The larger the amount of money left to be paid, the more incentive there is for customers to find ways to try to keep that money to themselves. At Smarter Remodeling, we leave no more than 2% to be collected at the end. We collect everything else before we get to that point. We even have a milestone in our contracts for punch out completion (which should be the stage where you get the job as close to completed as possible), leaving only 1 or 2% to be collected when the job is 100% completed (final inspections passed, permits closed, lien releases issued, etc.). My advice is to not leave more than 5% to be collected between punch out and final payment. By doing this, you keep control of the payment schedule and don't have to wait on the customer to receive a good part of your profits.

While I am a fervent proponent of staying ahead of the job regarding payments (collecting more than the work you have performed so far), I need to warn you about the risks of commingling this money with your other funds. Money you haven't earned yet should be kept in a separate account until such time your percentage of completion of the job has reached the point where that money now belongs to you. In simple terms, if you have collected 70% of the money for the job but only completed 50%, then set aside the extra 20% until such time you have completed 70% or more of the work. If you do this, then you should always be on a solid financial footing. If you start spending money ahead of completing the work, it will be just a matter of time before you regret it.

Some contractors (and most customers) think you shouldn't collect more money than the work you have performed. They think you should have enough capital to carry your business while waiting for payments. After all, if the customer doesn't pay you then you could place a lien on their property and collect what's owed to you.

I completely and wholeheartedly disagree.

Performing more work than you have gotten paid for is effectively using your money to finance your customer's project. This breaks the cardinal rule of *never* using your money to build a customer's project. It's asking for trouble and providing the customer with the incentive of finding ways to withhold the final payment.

If you have to place a lien on a property to get paid, now you're losing money two ways: first, you're losing the money you spent on the project; and second, you're losing the money you will spend on placing and enforcing the lien. This is on top of the amount of time you will spend trying to enforce the lien. Keyword here: trying. Why? Because placing a lien on a property doesn't guarantee you will get paid. Even if you enforce it (meaning take the process all the way to foreclosure on the property), you are not first in line to get paid. First the mortgage will get paid, then taxes owed will get paid, then anybody who had recorded a lien prior to you will get paid, and finally, if there is anything left, you will get paid. In other words, unless the homeowner owns the property free and clear or has a lot of equity on it, your chances of getting paid through a lien are very slim.

Use your capital to finance your business, not your customers' projects.

TRADITIONAL FINANCING

If you need money to finance your business, look first to your bank. Banks will almost always be the cheapest financing option, but if you are new in business or don't have good tax returns showing profits, then you won't qualify for bank loans and will have to go to the secondary market. This market (which has flourished in recent years) has many online lenders with daily and weekly payment loans. Cash from secondary markets are very easy to access as they base their approvals on cash flow rather than your credit or tax returns. However, their disadvantage is that they are very expensive and paying them back could use up all your profits.

I highly recommend you to use traditional financing, but if you must use a secondary lender, choose a loan that has no prepayment penalty and whose repayment plan won't hinder your cash flow.

Not showing profits in your tax return to minimize your tax liabilities may sound like an attractive proposition, but it will come back to haunt you when trying to obtain financing for your business to use for working capital or purchasing equipment or real estate. While using an appropriate, well designed, and especially a fully legal strategy (such as contributions to a 401K and purchasing depreciable equipment) to minimize tax exposure, it is highly advisable you demonstrate profitability in your tax returns if you want to pursue traditional financing. If financing is not on the table, then no worries: start collecting all those tax deductions. I'm fairly certain that a receipt from the strip club in Vegas won't pass as a meal and entertainment deduction, but, as always, consult with your tax or legal advisors before assuming something is a legal deduction.

SECONDARY MARKETS

At the date of the publication of this book there are two online lenders who I particularly like—neither of which I'm endorsing—but I can see value in their propositions.

The first online lender is Behalf. They are a financing company who grants a credit line so you can pay subcontractors, vendors, and suppliers. You apply to use their services and once approved you can start paying invoices with their credit line. They offer a few different repayment methods: 30 days with no fee to you (they will charge your vendor a fee if the vendor chooses to get paid by ACH instead of a credit card), then 12 weeks or 24 weeks. The longer the repayment period the more they charge you and they become as expensive as other online lenders. The reason I like them is because of the 30-day no fee option. I use them quite regularly to pay suppliers who don't mind getting paid with a credit card. When all is said and done, it allows me to gain an extra grace period of 30 days at no additional charge. If cash flow is tight and you want to avoid drawing from a credit line with a fee, then this is an excellent option.

The second online lender is Fundbox. They are another lender who will approve you for a certain credit line and who connects to your accounting software (QuickBooks and others). By doing this, they are able to see your account receivables (the invoices your customers haven't paid yet) and can offer to advance you that money, which you can repay in 6, 12, or 24 weeks. Their fee is not cheap, but if you find yourself in a situation where a payment is delayed because of a minor issue or when a customer is paying on terms (say 30 days out—at which time, you might want to include a fee in all your invoices so that customer will cover the fee you will need to pay in order to get the money advanced to you), you can get the money from Fundbox to avoid cash flow issues. Once your customer pays, you can pay Fundbox back early and save on the fees.

Both alternatives need to be analyzed carefully (read the fine print!) and you need to make sure they work for you and your business, but they are two of the best options in the secondary market.

To close this chapter, remember that cash flow can make or break your business as easily as running jobs with little profits can, so plan ahead. Establish credit lines with banks or alternative lenders, even if you don't use them, because you'll never know when you might need them and it's always better to have them and not use them than need them and not have them.

Lesson of Success

— 41 —

Preplan and prepare for financing.

Regardless of your opinions of using banks or lenders to support your cash flow, it's better to have them lined up for an emergency situation than it is to be unprepared.

Set up your credit lines. Most credit options don't require you to use them, but do read the fine print, as occasionally some plans do require that they be used to keep them active.

Lesson of Success

— 42 —

Dictate customer payment terms.

This is your business. Why would you let a customer determine the way you want to run it? Never allow a customer to dictate the payment terms, much less allow a customer to get ahead of the game by performing more work than for how much money you have collected. This is your show, so get paid to perform.

CHAPTER 17

TEAMS—THE ORGANIZATIONAL CHART

Every successful remodeling business needs an organizational chart outlining all the positions that need to be filled for the company to run smoothly, profitably, and efficiently. And one more thing, the most important one: it has to run without you, something the organization chart should demonstrate.

This is critical to building a business (and not a job) which will allow you to spend more time doing the things you love instead of the things you must do. This doesn't mean abandoning your business, only that it doesn't require as much of your time.

Once you create an organizational chart, your name will probably be in most of the boxes. As most entrepreneurs, initially you will wear most of the hats in the company. In the early days of Smarter Remodeling I was the "hat-wearer" and images of the Mad Hatter come to mind. It always seemed that no matter how many hours I worked, I still needed a few more.

The critical reason for building an organizational chart— and to begin working in filling those positions— is so that you can be relieved of certain burdens and can concentrate on growing your business. I used one of our networking contacts (who is a business coach) to help us create the organizational chart for Smarter Remodeling and found his advice so valuable that he is now a mentor for my new venture: SR360 Solutions.

Don't be afraid to look for help, as you are not going to build a successful business without some of it. However, get the right help, and stay away from anything that looks or smells like a "get rich quick" scheme.

Rather than put up a formal, deep chart, here is a simple process to follow

1. List yourself at the top.

2. If you are planning to be an eight-person company, create three categories to fill the eight spaces below.

3. There are three main functions in your type of business: office staff (admin, payroll), field workers (carpenters, general laborers), and support staff (sales, managers, assistants).

Under your name should be three boxes with the functions or categories: office, workers, and support. Once you hire a specific person to fill that function, you can give that person their own 'block'.

Under each of those categories, list the people by name, remembering that in the future you want to see your company as an eight-person company.

	Boss *Tony* Office	
Office	Workers	Support
General Office *Sally*	Crew Manager *Tom*	Sales *Bill*
Payroll *Tony*	Carpenter *Alan*	Assistant *Jack*
	Trim Carpenter *Sam*	Project Manager *Tony*

This is where company organization starts and it can continue to be expanded as your business grows.

After you start filling those boxes in the organizational chart with names other than yours, things get fun. Then it's time to concentrate in creating systems and procedures that are repeatable, ensuring continuous profits, keeping your team engaged, and, most importantly, allowing your business to run without you.

Creating systems and procedures may sound ominous and you might be thinking, how in the heck do I do that? But it is much simpler than you think. After all, you do it every day; you just didn't have a name for it. But, more on that in our next chapter.

Lesson of Success

— 43 —

Create your dream organization chart.

Spend time thinking about what positions your dream company will have. In the beginning you probably will have your name in most boxes but over time, this chart will help you hire the right people who fit into the right boxes. A business coach would be a great person to ask for assistance to complete this task.

SCAN ME FOR MORE CONTENT

CHAPTER 18

SYSTEMS AND PROCEDURES

If you are already running your business, there are already things you are doing right. There are things that work for you which are normally repeated with every job you do. This could be pricing, sales, or construction. Every one of these tasks can be converted into a system or procedure so new employees can learn the way you want the task done, and just like that, training has been taken off your plate so you can concentrate on … you got it … building your business!

A procedure is a single task, like how to measure cabinets. A system is a combination of procedures, such as building a kitchen: it utilizes the procedures of how to measure, schedule trades, complete selections, obtain permitting, etc. Each one of these tasks is a procedure, and the combination of them allow you to have a system to build a kitchen. The more systems you have in place, the easier it is to teach others. The more people you teach the less you have to do which in turn becomes your way of … you got it … building your business!

To exemplify the importance of systems, let me tell you a short story about my daughter, Karen.

When Karen was a little girl, she loved to have her milk frothed and my wife, Agata, would prepare it for her just so (little did we know that we were creating an addiction that would result in thousands of dollars spent at Starbucks!). Anyway, frothing the milk involved several steps (procedure) which produced beautifully frothed milk.

Karen wasn't a big fan of other dairy products, so it was important that she drank milk so she could get the vitamins and minerals she needed.

Agata had developed a series of procedures beginning with the volume of milk poured into her cup, then how long to froth it, how long it needed to be steamed, etc.

Every step was carefully completed, including the minute tasks of warming the cup correctly so that the froth would develop a certain way, ensuring that it would stand up longer. In short, Agata has developed a system to froth Karen's milk that has worked for years. She had begged me to read the instructions she had clearly written down, but since there were too many steps involved, I always avoided the milk frothing chore.

One day, Agata was doing a few errands and Karen asked me for frothed milk. I didn't want to do it, so I tried to feed her another snack, but she kept insisting for her milk.

Finally, I got tired of her constant whining and capitulated to her demands. Mumbling an indescribable promise for retribution as a repayment for her audacious milk request, I grabbed the milk jug, poured it into a cup, turned on the coffee machine to prepare the frother, and waited for a minute or two: I was missing my favorite TV show (yes, I'm old enough there was no TIVO or live pause yet).

Once the light on the frother warned me it was hot enough, I opened the valve and placed the cup under it—the steam was still going off. In a matter of seconds the kitchen was covered in milk, including myself … I wasn't happy.

For a fleeting moment I considered the possibility of blaming Karen for the mess and making her clean up the kitchen, but my fatherly (well, survival) instincts kicked in and warned me that the punishment coming from Agata would be far greater than the inconvenience of cleaning it up myself; so, I started cleaning.

I swear my little sweet daughter's chubby face was smirking beneath the hands that covered her face. I couldn't prove it, so there was nothing that could be done.

Once the kitchen was clean, I refilled the cup, this time placing the frother in the cup while slowly opening the valve.

I had wasted valuable show-watching time, so I rushed the frothing until the milk foam was just going over the rim of the cup. I sprinkled sugar on it and handed it to Karen, rushing to the TV just in time to realize they had revealed who the killer was but weren't repeating it … NOOOO!!

Now who was it?

I was in the process of considering all the characters in the show who could be the killer when I heard a scream from the kitchen. A visceral, penetrating scream that tore into my ears.

I jumped off the couch like I was a member of the USA Olympic gymnastics team and ran to the kitchen. Karen was staring at the cup with tears in her eyes while Agata, who had just walked in, ran towards her baby, laser focusing all her witch powers on me. I was astonished to see her home while growing concerned for Karen who kept screaming, "Burn, burn."

I'm dead, I thought.

The damn milk was too hot and I burned Karen's tongue. I could already imagine Agata hanging me head-down on the pool deck, covered in honey so the ants would eat me alive.

Both Agata and I reached Karen at the same time—which by itself was a miracle of physics because I was two steps away and Agata was several steps—but mothers are known to bend the laws of physics for their children.

Agata grabbed Karen and started blowing in her mouth but Karen shook her head and pointed to the cup.

I asked, "What? Too hot?"

Karen looked at me with malice, knowing she was about to get me in trouble, and with puppy eyes looked at Agata and said, "Daddy burned the milk!"

Agata grabbed the cup and, like a milk connoisseur, smelled it, looked at it, and confirmed the diagnosis. The milk was effectively burnt.

Agata looked at me and very clearly said, "There is a piece of paper with step-by-step instructions on how to froth Karen's milk. You didn't see it? I have shown it to you many times."

Karen was the only one enjoying the moment, and she smiled sweetly while I was reprimanded. I promised I would never do it again: I would stick to the systems from then on. And I did. I never burned the milk again (mostly because I never made it again!) but I did learn the importance of creating and following procedures and systems, which have made all my companies efficient, profitable, and simple to run.

Lesson of Success

— 44 —

Establish procedures and systems.

Create the procedures and systems so everybody in your company knows what to do at all times. I know it sounds like something only large companies do but they are necessary if you intend on building a company with one or more employees.

Creating procedures and systems is a simple thing to do. Write down *how* you do things and *why,* so new employees can learn and do the same.

Start with something small. For example, write the procedure on how to open your office (alarms code, tricks with the lock (if any), where the lights are, etc.). Then write another. And then, another. Before you know it you will have many procedures on how to run your business.

CHAPTER 19

LEARNING—CONTINUING EDUCATION

Never stop learning. Always set time aside to learn. Read or listen to a book. Attend conferences and tradeshows. Learning inspires you, which allows you to inspire your team.

Personally, I don't have much time to read, so I've started listening to audiobooks while driving. Over the years, audiobooks have allowed me to listen to hundreds of books, thereby giving myself the opportunity to learn from incredible people around the world. There is always an idea floating in the ether that can help your business or yourself personally. Never miss an opportunity to learn and always be open to it.

Even as we are in the process of launching a nationwide software business which will empower thousands of construction business owners to run better remodeling businesses, I'm still searching for new ideas so I can improve and help others to improve and be more profitable remodelers.

This concept of helping others is what eventually would become the Smarter Remodeling Skilled Trades Network™ and the Smarter Remodeling Construction Academy™.

Both initiatives were created as a way to create differentiators with our competition, but most importantly, they arose out of the belief that through others' success we will become successful too.

*****

Over 20 years ago, when we started Smarter Remodeling, we did most of the work ourselves, and slowly over time we began hiring employees and working with subcontractors. Always finding the right employee was extremely difficult, and making sure they performed at high standards was even harder. We started working with subcontractors, but customers didn't like that much. I didn't understand why so I asked.

"Well," one customer told me, "They are not your employees so you cannot dictate the quality of their work nor the hours they work nor how they work. We hired you because we liked *you* and now, we have strangers working in our home."

It never crossed my mind that customers felt that way about subcontractors. I stopped hiring subcontractors and instead focused on hiring more employees. But my profits were going down and my headaches went up. I had to find a way to use subcontractors or I wouldn't be able to keep growing.

After spending time with one of my subcontractors, I learned about his challenges. I realized how easy it was for him to complete jobs but how difficult it was for him to run the business side of things, especially since he wasn't a businessman.

He had started his own business because of differences with his boss and he wanted to set his own hours and make more money.

I told the subcontractor, "Those are great goals and you should work hard to achieve them."

He looked at me with disappointment and replied, "I work hard. In fact, I work harder than I have ever worked before but I still can't finish everything I need to do in the business. I can't find the hours in a day to complete it all. I don't know how much longer I can continue doing this. I was seriously considering asking you for a job."

I listened intently, absorbing every nuance of his voice and realizing he was feeling defeated by the business he had given everything to.

I put my hand on his shoulder and said, "What about if I can give you something better than a job?"

He looked surprised but hopeful.

"How about if I teach you how to run your business in a profitable way that gives you more time to spend doing the things you love to do?" I asked.

His eyes misted. "Is that possible? Why would you do that? What's in it for you?" he asked.

I smiled. "Seeing you succeed is reward enough; but I'm convinced that helping you stay in business is also good for me, for my business. Let me try to help you. You have nothing to lose."

He stretched out his hand, clasping mine. "I'll be forever grateful. Let's get started."

Just like that I had my first trade partner. I didn't know it at the time, but this would evolve into a one-of-a-kind subcontractors' network. It became so successful that at some point we had to stop accepting applications for trade partners.

I taught the subcontractors how to properly price a job, even when it goes against the cardinal rule of contractors: always get the cheapest price possible from your subcontractors. I taught them how to use accounting software; I connected them with insurance agents and bookkeepers. I helped them set up automatic payroll systems. From the smallest to the largest of tasks, I helped many of them succeed.

The first trades that completed the training were accepted to be part of our network. They had to take a separate training course to understand our company's values, to learn how we wanted jobs done, and to learn how we treated our customers. The program was not only good for us but for them as well, as the principles instilled could be applied with any customer, not just ours. The level of synergy we received from our trades was such that the difference between employees and subcontractors was no longer present.

Throughout the program, we removed one of our customers' perceived weaknesses about the quality and trustworthiness of subcontractors and converted it into a strength.

We also received the added benefit of not having a revolving door of trades but a stable, trustworthy, value-oriented network of skilled tradespeople available to serve any and all of our customers.

Fast forward to today, and we have added the Construction Academy, where we bring in tradespeople to perfect their trade with continuing education, enhancing their skills, teaching new techniques, and updating them on code compliance. The best part for the tradespeople is that it's a free service to anyone who is part of our Skilled Trade Network™.

*****

Nowadays I feel great looking at the past and knowing I was right about the importance of nurturing relationships and investing in the success of others.

We have invested hundreds of thousands of dollars over the years to help trades acquire equipment, technology, and more. If I would have had a network like this at my disposal when I started, I'd have saved tons of money and headaches … and more than one heartbreak.

Finally, by teaching others I learned a lot of new things that I was able to apply in my business, making it better, more resilient to changes, and especially, more profitable.

Lesson of Success

— 45 —

Always be learning.

You don't have to be a student to learn. You can be a teacher and still learn lots of new things, from the material you prepared to the conversations that develop with your students or peers.

I've been surprised with the ideas that others have come up with when asked about certain situations.

There is always an opportunity to learn if you are open and willing to. You can learn from your mistakes, you can get ideas from books or audiobooks, from watching a video, and from discussions with others.

There is always an opportunity to learn, so never stop looking for one. You never know where your inspiration for your next great idea will come from.

CHAPTER 20

NETWORKING—BUILDING RELATIONSHIPS

I don't have enough words to express how important it is to invest in building relationships. When I use the term *invest,* I'm not only referring to money but also to time, emotional capital, and skills.

The amount of business I receive—without actively seeking it and is solely based on the relationships I have created throughout my career—exceeds 30% of our total revenues every year. It is by far the best return on investment one can make. For the most part, these relationships only require investing time but the key is, you genuinely need to care about the other party: listen to their concerns and see how you can help. Perhaps you can't help them but know somebody who can.

I've been part of a networking organization called BNI for the past 7 or so years, and I have estimated a return of over 10,000% on what I spent on dues. Not only that, but I have gotten to know people I now consider friends, some of whom are part of my team of advisors.

When you actively network the possibilities are unlimited. Networking changes from working with leads to working with referrals. Suddenly there are people who are expecting your call and to whom you are no longer a stranger: you have been referred to by somebody they know.

Look for networking groups in your area and join one. BNI is a paid networking group and its the largest of its kind, but there are others that are free to join if your funds are limited. The most important aspect to successful networking is to show up consistently and actively participate in the meetings. Be committed and dedicated to your meetings and fellow networkers. Get to know your fellow networkers. You must interact with them and get to know what kind of business they are after. Set up dates for after the meetings so you can do individual interviews with each member and learn about their business as they learn about yours.

There is no such thing as passive networking. You must invest time in building relationships and make the effort to know other people so they start trusting you and begin referring business to you. Don't expect to show up to a meeting and walk away with tons of business. It's possible, but highly unlikely. However, if you keep showing up and participating, you will get some referrals and you will close some business.

Lesson of Success
— 46 —

Learn to network.

At the beginning of my career, I thought networking was an opportunity for me to sell to the people I met. I was mistaken. If you network with 20 people and your only goal is to sell them a product or service, then you will be done quickly as your pool of customers is very limited. But, if you invest your time in getting to know them, educating them on your business and the class of referrals you are looking for, you will create a salesforce of 20

---

people who can bring you more business than you ever dream of. Like a saying in BNI goes, "Networking is more about farming than hunting."

CHAPTER 21

THE IMPORTANCE OF PROFITS— P3

My business philosophy has always been P3: Profits, Profits, and Profits. Profits are the foundation of a solid and successful business that provides incredible customer service to its clients, a great place to work for its employees, a generous community partner, and solid returns for its shareholders. Many people misunderstand this; they believe this philosophy is based on greed. On the contrary, this philosophy is based on the reality that a business without profits can't survive and therefore can't provide anything to anybody.

The P3 philosophy entails creating value for all the company stakeholders: clients, employees, shareholders, and the communities where we do business. Profits are an intrinsic core value for us because we cannot function without them and our customers cannot rely on us if we'll soon be out of business. At Smarter Remodeling, if we don't have profits then we can't fulfill our corporate goal of being an active community partner. We provide access to free remodeling services to people who cannot afford them through our Smarter Remodeling Cares initiative.

For many people, profit is a dirty word. It is commonly associated with corporate greed and lavish lifestyles built on the back of workers. For many, just the mention that a company needs profits to survive and thrive is a synonym for horrible excuses to fatten the business owner's wallets. This misguided and preconceived notion is fed by movies like Wall Street and makes it difficult for most contractors to be profitable rather than barely surviving.

For many contractors it's extremely difficult to justify the markups they require to make a healthy profit. It's this culture that permeates to all levels of the organization, making it extremely difficult to achieve profitability, and almost ensuring the company will eventually fail and go out of business. Because it's difficult to determine what the right markup is, you need to turn a profit; you first need to understand and determine what a healthy profit means to you.

For most construction companies, a net profit of 8-10% is considered healthy. No, this doesn't mean you simply add 10% to your costs and you're set to go. Nothing is further from reality than that. In order to achieve a 8-10% *net* profit, you first need to account for all your overhead expenses and job costs. We discussed this in detail in Chapter 13: Profitable Pricing.

You need to make profits the engine that propels your entire organization. My concept revolves around making profits a core value of your organization. My belief is that profits should be at the heart of the organization, front and center. All systems and procedures should be planned to maximize profits.

I know this sounds greedy, but take a few moments to reflect on it. How can you provide the best customer service your clients have ever experienced if you don't make any money to do so? You can't. Without customer service you will always be in survival mode, looking for the next client, which will take precedence over providing great customer service to existing or past customers.

How do you retain the best employees if you can't afford to pay them? Or how do you come up with the ideas to further your business if you can't afford the time to bring them to life? Being profitable will allow you to do all that and to do it with a smile on your face.

Profits are critical to any business and you should spend time thinking about ways to maximize them. Once you have done that, do it again.

Teach your employees to think about profits and why. I guarantee that they are not working just for fun. They need to make money to pay their bills; they need to be appreciated and feel a sense of accomplishment; and they need benefits and perks. But you can't afford to give employees any of that unless your business is profitable. You need to teach them *why* they should be focused on making profits for the company and that, as explained above, is because those profits allow them to provide a high level of service that makes customers happy—no one wants to be yelled at and everyone wants to enjoy their job!

Profits pay for bonuses, vacation time, and benefits. For decades, business owners have been afraid to speak about profits with their employees as they often believe if the employees see how much money the company is making, then they would want more. While this may be true in some capacity, it's your job as the leader to explain why the company needs the profits it makes: to pay for employee benefits and for customer service—and what you get as the owner and risk taker.

Don't be afraid to put the numbers on the table. I have learned that most employees understand much better why we charge our prices once they see the numbers under "the hood."

Put a plan in place to share profits with your employees. Whatever profit they generate above and beyond your standard markup, you should find a way to share that "wealth" with them. This provides them clear incentives to maximize profits as they are getting a benefit from that as well.

The year 2021 was a banner year for Smarter Remodeling. After having a great year during the pandemic year of 2020, we started 2021 with more doubts than certainties. Fortunately, we were able to focus on our core strengths and closed 2021 with revenues 20% higher than our best year.

Using the criteria for recruitment we had created years earlier with our organizational chart and other processes, we grew our staff by 40%, thereby putting us in a great position to keep growing from year to year.

The software we developed for ourselves and potentially for many other construction business owners across the country changed the way we approach our daily business activities. It is allowing us to focus more on areas of profitability while increasing our return on investment on our marketing dollars.

At our company Christmas party, we had over 50 attendees and had presents for everybody, including 70-inch TVs, appliances, and Apple AirPods. We handed out Christmas bonuses that brought some employees to tears and signed a contract to assist a non-profit foundation to build two houses for low-income families.

All of this reminded me why we worked so hard in instilling a culture where profits were put front and center and they were used in the service of employees, customers, and our community. We do make a difference in people's lives.

In our new company, Smarter Remodeling 360 Solutions, our mission is to make every contractor a profit-oriented company to benefit all its stakeholders. By focusing on company stakeholders, we can create enterprises that are socially responsible and ethical, focusing on providing value to its communities.

True to our values, we can help "remodel" this country to once again be a beacon for entrepreneurs around the world.

Lesson of Success

— 47 —

Think about profits.

Spend some time thinking about profits and the numbers your company need to achieve its mission; but don't make the mistake of only raising prices to achieve that number. With each price increase you need to provide an equal or greater amount of value so your customers can perceive they are still getting a deal.

Many contractors believe they can raise prices and customers will pay up. While this might be the case, they might also end up paying somebody else who could provide them with greater value.

Sometimes the solution is not in increasing prices but rather making your company more efficient and reducing overhead costs.

CHAPTER 22

---◗◆◖---

FREE ESTIMATES

One false notion most people have is that free estimates are the way to get business. While this model might work for many, I can almost guarantee you will be stuck in a vicious cycle of estimates if you give in to this.

I cringe when I hear contractors speak of their plans to provide free estimates to grow their business. It's as if their time and experience are worth nothing.

If you don't put value in yourself and your company, nobody will. Everyone will perceive your time has no value and you are confirming their assumptions. The free estimate model may work for companies who don't put in real time or effort to create an estimate. Their numbers work by just multiplying quantities by values and that's it. But for everyone else who gathers quotes by spending time, measuring, designing, and consulting, they need to charge for that time.

I have no idea when the free estimate craze began but I wish I could find a DeLorean to go back in time and stop it … and maybe leave myself a sports almanac too! Whoever came up with the idea did the industry a huge disservice. They set up all future contractors to spend lots of time running around doing nothing but giving people free estimates, most of which will never result in a job.

It is mind-boggling that people think contractors should use the wealth of information they have learned over the years to give away free estimates. Many customers have no intention of buying their services and since the estimate is free, they call contractors for prices on the latest dream project they saw on HGTV.

I have talked with customers who were convinced contractors were obligated to provide free estimates. They stopped short of saying it was mandated by law but it was implied.

I was one of those contractors. In the early years at Smarter Remodeling it was just myself running around doing all the free estimates and it was very time consuming. In addition to the estimating, I needed to have time to source materials, coordinate projects, do some business paperwork, etc. Since I was new in the business, I pursued every lead I got, so I only had time to write estimates ... and for free!

I finally got smarter and started qualifying the leads I got, declining many of them before I ever set foot on the property: they were too small, too large, or not the kind of work I wanted to do or could profitably do.

My first lesson was never to run a lead without first doing pre-qualification work. Here is where different authors and construction business owners have varying opinions: some believe in chasing every lead and running your qualification process in person, while others say to pre-qualify on the phone.

I'm inclined to suggest a middle ground. Begin by running a soft pre-qualification on the phone, scanning for potential or red flags, but complete the qualification in person, and demonstrate to potential customers the value you bring to the job. Once your value is established, proceed to present a list of services which requires the customer to pay to get access to. One of the services that should cost money is a written proposal which includes a detailed scope of work.

A detailed proposal differs greatly from a price range or ballpark estimate. These are general ideas of prices for projects that resemble (or don't resemble) the project your customer wants.

A ballpark estimate carries a great deal of uncertainty for both you and the customer, as there is no clear and well-defined scope of the project. Ballpark figures lack the ability to define and determine what the customer is after: a modest remodel or a top-of-the-line, brand new creation. This is why customers need to invest in getting a written estimate that's specific for their project with real and specific numbers.

To produce specific numbers you need to invest time, make use of your experience, and consult with the network of subcontractors which you've built over the course of many years.

But *why* should you charge for estimates? Well, the answer is very simple: if you don't value your time, how do you expect your customer to? Something that's free has no inherent value, so if your time to produce an estimate is free then you are starting that relationship with the wrong tone. They will expect more things from you, for free.

Once you have discussed numbers with your customer and both understand the scope of the project, then charging for an estimate allows you to separate the "just looking for a quote" kind of customers from the ones that are really serious and ready to start. Time is finite, so which potential customer should you concentrate on? The paying customer.

When Smarter Remodeling began charging for estimates, I had an interaction with a customer that still makes me laugh at how some people assign automatic value to some professions but not to others. It's always a matter of perception.

I had stopped to look at a potential customer property to whom I was referred by a friend. Because of this, the customer hasn't passed through the typical pre-qualification process (which taught me that all new customers should go through it, regardless of who referred them).

I introduced myself and after a few pleasantries, the lady began showing the addition she wanted to get done. We talked about her needs, her wants, and her wishes for the addition and the many reasons why the addition made sense. I asked a few times about how much she was comfortable investing in a project like this; she skirted around the questions without providing a clear answer.

After another ten minutes of conversation, I told her a project like the one discussed would be in the range of $150,000 to $250,000, depending on the final dimensions, finishes used, etc. She looked shocked at first but still asked me to produce a detailed estimate and some preliminary drawings for her to see and further evaluate the project.

I was happy to do so in exchange for a $1,500 fee. The fee would include a detailed estimate including allowances for materials and a basic layout of the project for her to visualize the project. I stated that it would take ten to fourteen days to complete as there was a lot of information I needed to collect from subcontractors and vendors in addition to the time needed to produce the drawings.

She looked at me as I had suddenly grown two heads. "Excuse me?" she said. "Nobody charges for giving prices. Are you trying to charge me for a free estimate?"

Silently laughing, I looked at her with compassion and replied, "No, ma'am. I'm not charging you for a free estimate. I'm charging you for a detailed proposal, and that's not free. I gave you a free estimate already, but we've agreed that you want a price that reflects the exact scope of work for your project, including layout drawings, and that is what I'm charging you for."

She was getting more upset by the minute. "This is unbelievable. I don't understand why I would need to pay for this when everyone else does it for free," she said.

I nodded my head and proceeded to explain. "I'm aware that many contractors use a free estimate business model to attract customers; however, it's simply something we cannot afford to do. We have spent a better part of two decades educating ourselves and acquiring the needed experience to produce this kind of proposal. A lot of money was spent in the process so we could offer, not just an estimate, but an accurate and detailed proposal to complete a project like the one you're looking for. We are true professionals, and as such we hold ourselves to a much higher standard than other contractors, which is why we can offer an On Time On Budget Guarantee, which basically means we will build your project for the price we quoted within the time we estimated. Or we will pay you!"

I gave her a big smile to wrap my speech but realized it had not produced the expected result.

She smirked and said, "My son is a doctor. He educated himself for over ten years and spent lots of money to get his degree. That's a real professional. Construction people are just trades." Ouch! That stung a lot.

I looked her in the eye, not ready to end the meeting yet, and told her, "I can understand your point, although professionals are not only people with medical or post-graduate degrees. Furthermore, I'm sure your son doesn't offer free consultations?" I didn't let her respond and continued. "If I go to see him and he gives me a diagnosis I don't like, can I just walk away without paying? Of course not. I can't remember the last time I visited a doctor and didn't have to produce insurance and payment information before seeing the doctor."

I took a breath and she took the opportunity to show me to the door. I gladly followed.

While we walked to the door, she breathed heavily, and was upset with me. "I can't believe you think you are at the same professional level as my son, a doctor. Unreal."

I turned on my heels like I'd been in a military honor guard my entire life and intently looked at her. "Of course not, ma'am." I paused for effect. "I'm sure your son couldn't build the addition you want. Have a good day."

I walked away holding my head high and filled with pride at being a professional contractor.

I didn't get the job and got an earful from the friend who had referred me to the job. While it wasn't the best situation, it afforded me the opportunity to defend my profession.

Perhaps, if more contractors start believing they deserve to be fairly compensated for their time and charge for estimates, then we could change the perception that contractors work for free and be recognized for the vital role we provide to our communities and the economy in general.

Lesson of Success

— 48 —

Establish a pre-qualification process.

You need to establish a qualification process for every potential customer that comes your way. Determine the best method for you: a phone qualification, an in-person, or a mix like the one Smarter Remodeling uses.

However, the following are critical:

1) Have a qualification process.

2) Charge for producing detailed proposals.

3) Don't give experience and time away for free.

SCAN ME FOR MORE CONTENT

CHAPTER 23

POTENTIAL CUSTOMERS

Not every customer is your customer. While every person might be a customer, they might not be *your* customer.

Don't be the answer to everyone's problem. You cannot serve every potential customer so why keep chasing work you might not want? In other words, if your goal is to build additions and kitchens, why chase leads for handyman work? Why waste time talking to customers who can't afford your services?

I partially agree with others who suggest not disqualifying people on the phone with a quick estimate to determine if the customer's budget is in line with the real cost of the project.

I partially agree because the customer may only have $5,000 to spend but the project they want is worth $20,000. However, if their budget is half or three-quarters of what you think the project would cost, it merits a brief site visit to introduce yourself, present the benefits of working with your company, showing examples of past projects, and discussing a realistic budget.

Price without context has little value. Customers cannot assign a value to a price, but they can determine if there is value in your proposition when they see what you and your company can offer.

Smarter Remodeling does charge for detailed proposals and design consultations, but we also offer free appointments based on the project, customer budget, and other variables.

On an initial visit, we don't spend much time with the customer, nor will we provide a design consultation or offer ideas on how to improve the project. That is a service we provide once we have been paid to do so. From the first visit we are showing the customer all the added value we have as a company and why we would be the smarter decision for their project. In other words, we are trying to sell ourselves to the customer, not a particular project.

Once we establish why we are a better choice than cheaper contractors, we discuss budgets and what a realistic budget for their project is; if they decide they can work with the new numbers, we proceed to sell our design services, where we produce a detailed proposal with guaranteed pricing.

We close approximately 50% of the leads we would have disqualified on the phone by visiting a site. It's well worth the effort, especially if you have salespeople who are hungry for qualified leads: let them convert a semi-qualified lead into a qualified one.

Potential leads don't need to waste much of your time, only enough to convey the value your company can add to their project.

This is not to say you should do free proposals. This is a process for you to put numbers into perspective with the value you provide.

Lesson of Success

— 49 —

Don't be hasty to disqualify potential customers.

Disqualifying customers on the phone without a pre-qualification process is a waste of advertising and marketing dollars.

Have a pre-qualification script to determine if this is a customer you may want to do a quick site visit or assign to a salesperson.

At Smarter Remodeling we charge for design consultations and detailed proposals but the ultimate goal is to qualify the potential customer; it is not to sell design services on a phone call. Have a qualification process and stick to it.

SCAN ME FOR MORE CONTENT

CHAPTER 24

———◆———

DON'T WORK FOR LESS THAN YOU DESERVE

Have you been offered work from a client who asks for discounts when given the price? Often these clients use the argument that they have more work for you in the future so it's in your best interest to give them a big discount now (or to even do the job for free).

That's a hard pass.

Never discount prices based on future promises. Why would you discount your prices in the first place? You are not Walmart with everyday low prices. Prices should be fair and reflect the value provided for your customers. If they need a discount to buy from you, then you haven't done a good job in communicating the value they are getting from you. I've seen this too many times and it never ends well for companies who use discounts to attract their customers.

Attaining customers based on price alone is just the prelude to you closing your business. I have repeated this conversation with many customers over the years. I played the discount game when I first went into business, but soon realized the customers I had acquired weren't willing to pay full price for other services.

They all expected discounted prices *on everything* which quickly ate into my profit margins.

Soon enough I found myself struggling to pay bills and working overtime hours to keep afloat. The situation changed when I realized reducing my prices to gain customers wasn't the answer to my problems. I started to build the value my company offered to customers, which kept my prices at a fair level.

I'm all for promotional offers based on value to customers, but not for offers that discount your price. When running a promotion, offer more value (creatively without increasing your costs) but keep the price at a level that allows you to maintain your profit margins. With every discount you give, you are losing a portion of your profits. It's like the silent partner who takes and takes but never contributes. It is bad enough that Uncle Sam with his tax collectors is a silent partner like this, no need to have a second one!

For the sake of clarity, let me repeat: always *add value* rather than discounting your price. Let me give an example.

You measured the house a customer wanted painted. You calculated the time it would take you to do so, listed all the materials you will use, added the equipment you need for the job, and proceeded to add your overhead and profits (markup). You determined your price to the customer is $9,500 and that you will pocket $4,000 after costs and overhead are covered.

You present the proposal to the customer who looks at you as though you had handed them a proposal to restore the Sistine Chapel. Then the diatribe to bring your price down starts. "What?! Are you sure? It's not that much work. I could do this myself in a couple of days."

This is the part when inevitably you consider telling the customer to go ahead and do it themselves, but you bite your tongue.

The tirade continues. "Come on, Mr. Contractor. I know you can do better. I understand you need to make a profit but this is excessive."

Have you ever wondered how customers can magically determine you are making a killing on the job? It took you a long time to determine the appropriate markup and what it takes to complete a job with a profit that enables you to stay in business; but somehow the customer can always determine how much that is in seconds.

Regrettably, this is when most contractors start doubting themselves. They can almost smell the fresh ink of a signed contract and almost every time they will cave: "Ok. Mr. Homeowner. What do you think it's a fair price for this job?"

And here your profits stay in Mr. Homeowner's bank account—and they are not coming back to you.

The homeowner doesn't know what a fair price is. Only you know the price needed to be charged to make a profit. The customer wanted a discount and now they know it's attainable. Whatever they can get out of you, it will be more money they keep in their wallet.

When a customer is asking you for a discount, they are asking you to take money out of *your* pocket and put it in *theirs*. A better way to respond when a customer asks you to reduce the price is to answer in the lines of, "I understand your concern with price and I'm sympathetic to your plight. Since I definitely would love to work with you, how about if instead of using Super Fantastic Paint Level 3, we only use Super Fantastic Paint Level 1. That could save you $500 and be more economical to you. The paint quality is not the same, so it won't last as long, but the price will be more to your liking."

You know the price difference from Level 3 to Level 1 is more like $800 but they don't. Watch how their face begins to change colors slightly. Their eyes start squinting as they try to get inside your head and deduce if you are playing them. Of course, they decline, "No, I want the Level 3 paint. That's the product I want to use."

You were expecting this so you reply, "Well, Mr Homeowner, I can't blame you. Level 3 is a superior product and I'd want to have it in my house, but I also can't afford it. So, what about if instead of two coats of paint I just do one coat and back roll it? This achieves a coverage very similar to two coats but, since it uses less product and time, I can offer you a price of $8,900. What do you think?"

If the right tone was used when you said you couldn't afford the paint either, then most likely you touched his ego. How dare you think Mr. Homeowner, possessor of the magic formula to determine contractors' profit margins in seconds, cannot afford your paltry services!? He looks at you after thinking for a few seconds and—so he can walk away with a story on how he defeated the greedy contractor—says "I want the two coats of Super Fantastic Paint Level 3. I still think your price is high but you will pressure wash the entire house, right? Because that's something I won't negotiate on." He closes his argument by crossing his arms.

You looked at him surprised as you have already explained three times that you will pressure wash the entire house to ensure the proper adhesion of the paint, so this is something you can agree to. "Of course, Mr. Homeowner. I wouldn't dare not to," you reply.

You stretch your arm, palm open to get a handshake that will close the deal while holding your notepad with the contract in your other hand.

You got to keep the profits in your wallet and Mr. Homeowner considered the pressure washing a win that he would convert into a battle of wits for his wife, in which he had to fight tooth and nail for you to cave to his demands.

Lesson of Success

— 50 —

Know how to negotiate discounts.

Don't be so quick to offer a discount to close a deal. Many times, customers just want something extra to feel they got the best part of the deal; but it doesn't mean you have to reach into your profits to give this. Be creative and be prepared.

CHAPTER 25

DON'T UNDERESTIMATE PEOPLE

Many contractors make the mistake of underestimating people. Often contractors want to talk to the decision makers, not to the employees. Why would they? The help does not make the decisions. What contractors often forget is you can get more information from the help. (Thanks to Kathryn Stockett for making the term 'The Help' widely popular.) You can use the information you get from them to your advantage and close a deal.

I have always had a demeanor that has allowed me to mix well with both kings and beggars (paraphrasing Rudyard Kipling now).

Remember from the beginning of this book my interactions with Mario, the shoe shiner. Since I grew up in a working-class family, I know being poor is not contagious, and I have no problems interacting with different kinds of people. I draw the line at Steelers and Boca Juniors fans though!

My ability to establish conversations at all levels has opened doors to closing large business deals, a skill every business person must learn.

Several years ago my company was on the verge of crossing a huge revenue threshold we had been pursuing for a while, but kept missing the mark year after year. I got wind that a large national company with a local presence was looking for a contractor to do work in their facility.

I was familiar with the facility, having been in it years before they bought it. I did not know who was in charge of the project and began putting feelers out in my networking groups that I was looking for a contact inside that company.

A few weeks later a networking buddy gave me the contact of somebody who worked there. His name was Pete Jax; he worked in maintenance. He didn't know much about Pete, but he was willing to call him and ask him to meet me.

I took the opportunity and my friend placed the call. Pete agreed to meet me at the company facility during his lunch hour. I made it on time and Pete walked me through security and back to a large parking lot where there were a few tables and chairs in the furthest corner from the building.

With no trees for shade, the unforgiving sun of a Florida summer was beating on us. As we got close to the tables, I asked him if there was a place to sit other than in the sweltering heat. I was more concerned about the sweat that had already started running down my back than business! Pete told me it was the designated place for janitors and maintenance to eat lunch.

I asked about the cafeteria we walked by on our way to the parking lot and was informed it was only for middle management and visitors. I smiled at Pete and said, "I guess we're in luck then because the badge security gave me says 'Visitor'. That entitles me to use that cafeteria."

He smiled and laughed, pointing back at the building. "I have never eaten lunch there since I always pack mine."

I wrapped my arm around his shoulder—which took him by surprise as he was wearing a dirty overall and I was wearing a white buttoned up company shirt—and said,

"Today is the day you get to eat there and take your lunch back home as today's lunch is on me."

He kept laughing, showing a smile with missing teeth, but it was clear he didn't care. We walked to the cafeteria and I guided him to the ordering line. The cafeteria staff and those already seated stared at him. I flashed my Visitor badge at them, and with my best impersonation of Leeloo in the Fifth Element movie said, "Multipass."

Pete laughed again but the others didn't. I told the servers Pete was with me and to serve whatever he wanted. Pete proceeded to order food … a lot of food and a couple sodas. I ordered one entree and one soda; paid for both and left a generous tip that elicited a forced smile from the line server.

I walked to a table and sat down. Pete kept smiling. I asked why he was so happy, after all this was just a corporate cafeteria, nothing fancy. He said he would have been shown the door if it wasn't for me being there, and since I was dressed like the others and had a badge, they couldn't make an issue about it: they didn't know how important, or not, I was.

It was my turn to laugh and I repeated, "Multipass." Pointing to my visitor's badge. He laughed again, loudly causing people to look in our direction.

Pete was really a great, humble guy. He had worked at the facility for years long before the new corporation bought the place and took over. He knew everybody there.

He started to point at people there and said who they were. The VP of Communications was seated by himself reading a book. The Facilities Manager was sitting with other managers. The VP of Construction was sitting with contractors. I stopped Pete and asked him to tell me more about that guy and the contractors who were sitting with him. He said the guy was in charge of all the construction projects and then named the contractors by name and company.

I was well aware of their companies. There was no chance in hell I would get a job here if the VP of Construction was talking to companies of that size: they were all 100 times bigger than we were.

Pete kept talking and I enjoyed the conversation as I already knew I wouldn't work there. He had a lot to say about everybody. It was obvious that even when other people didn't notice him, he had been noticing them, and taking notes. I was laughing at some of his remarks when he circled back to the VP of Construction.

"That guy doesn't know it yet but he is on the way out. There is a new guy here, Albert Cain. I overheard him telling an assistant he was taking over the area as soon as the merger went through. He also said he didn't want to deal with huge companies as they sucked away every dollar he had for larger projects, and he would rather do deferred maintenance than get into huge projects now."

I was mesmerized by Pete's talk. It was like he was singing to me, and every word was music to my ears. I asked where Albert was. Pete said he was almost never there as he was in charge of other facilities too, but he heard when he was in town he played golf at a place close by, as he couldn't get into 'The Players' course. I wasn't surprised. The Players is where one of the most renowned golf events in the world was played every year, and I could imagine access was either very restricted or very expensive, or both.

I was so happy with the information Pete provided me, I could have crossed the table and kissed him; but I held back. I told Pete I would consider it a personal favor if he could call me when he knew Albert would be at the facility and he promised he would.

Less than a week later Pete delivered on his promise. He called me to tell me Albert would be at the facility and he had talked with him, saying I was a great guy to know and it could benefit him to spare five minutes to talk to me. Albert obliged and was expecting me two days later. I was excited.

I told Pete I owed him one and he replied having lunch the other day was enough. He appreciated me taking the time to have a conversation with him, treating him like he was a person and not an extension of the vacuum cleaner, and listening to his stories. That resonated with me. I felt sorry for him, and I promised myself that I would take care of Pete if I could close a deal at the facility.

I made a few calls and was able to get a round of golf at the Sawgrass Players for four people—for almost $4,000! What the heck?! I knew a round of golf cost under $200 in most nice courses but this was like $1,000 a person. I was wondering if Tiger Woods would be carrying my clubs for that price, but mostly I was happy I got a tee time.

I showed up to our meeting ten minutes early and Albert came to security to collect me. He asked me about my company and I explained the services we could provide. As Pete had already told me what some of the deferred maintenance items were, I had run some numbers and was prepared.

Albert liked that we could provide those services, as he had been thinking about several maintenance items. He had replaced the VP of Construction and was trying to understand if he could fit all the maintenance needs into his budget, using the remaining budgeted money to take care of old business, which gave him more time to prepare for a larger upgrade later on.

I inquired about the scope of work and he gave me the details (Pete was 99% right), and then I asked him what the budget was for all those items.

"I don't need you to tell me how much you think those items will cost. Just tell me what you want to spend." I told Albert.

He stared at me for a few seconds and hesitantly told me. The number was almost double what I had calculated. I immediately assumed I had made a mistake in my calculations.

I looked at Albert incredulously. He jumped the gun and said, "I know it's not enough but it's all I have. If I could find someone to do all of those items for this price, that would go a long way for me to make my stripes in this place."

We discussed the scope of work and I came to the conclusion that I hadn't made any mistakes: rather I had overbudgeted a few items.

We discussed all the projects we had completed in other places, how we like to do business, and the value we can bring to companies like his. He was excited about it, but said as much as he wanted to work with us, he had to bid on all the parts of the scope so he knew what he could afford.

"I understand." I said, "This is like playing golf. Most people think you just grab a club and hit. They don't think about planning your shots carefully, to come out a winner. It's not luck; it's planning."

Albert grinned. "Exactly! You play golf?"

I shook my head, "No, I'm one of the people who gave meaning to that phrase. I just grab a club and hit ... or, at least I try." Albert laughed hard and I continued, "I'm mostly there to have a few beers with my friends; but most importantly, I drive the golf cart."

He softly punched my shoulder and said, "You are a riot!"

I shrugged.

"Seriously," I said. "I must have driven in most of the courses around here and hold the record for best driving and worst score in almost all of them."

He couldn't stop laughing, so I pushed on. "You play?"

He nodded affirmatively. "I try. But it's hard to get time in the good courses."

This was my opportunity. "I know," I said. "I have been able to get tee time to play this weekend at The Players. Do you know it?"

He was taken aback by the question, "Do I know it? It has been my dream to play there forever, but it's impossible to get time there unless you are Tiger Woods or really well connected. How did you get a tee time there?"

I smiled and got close, "I could tell you, but then I would have to kill you."

He laughed hard and I took the opportunity to ask, "Would you like to come? Two of my friends who were supposed to come are out of town and can't make it. I was planning to take my son, but I have two free spots. They are yours if you want them."

He looked like I said I'd bring Jack Nicklaus to mow his lawn.

"Are you serious?" he asked.

"Of course," I replied. "It would be my pleasure and it would also give me the opportunity to discuss your projects with my team and subcontractors. Maybe there is a way we can do all of them for your budget."

Albert was in heaven. He would play at his all-time favorite golf course and complete all the maintenance items on his list. This was better than Mariah Carey seated on his lap, singing Christmas carols.

"Fabian, if you can do that, I'm ready to sign a contract and make you the exclusive contractor of this facility," he said.

I offered my hand, "Then, let's get to work as I believe we can make this happen."

We closed the deal the week after playing at The Players, at which I once again got the worst record in existence. I brought my lead project manager and Albert brought his second in command. We got to know each other and signed a contract that pushed our revenue to exceed our mark by almost 20%. It opened the door to many projects in the years that followed.

I took good care of Pete, and I kept doing so over the years as I built a relationship with Albert.

Now when I go to the facility I use the executives' cafeteria with Albert, but every once in a while I find Pete and bring him there to show my appreciation and to remind him of the difference he made for the future of my company. If he ever wanted to leave, he would have a job with my company.

Pete hugs me every time he sees me, and more often than not he stains my shirt. I don't mind. I can take my shirts to the laundry and I have many others; but I don't have too many Petes. His friendship is priceless.

Lesson of Success
— 51 —

Don't underestimate the help.

Never underestimate the power people lower on the totem pole hold. Treat everybody with respect and learn to listen.

Knowing who your customers are, what they want, and what they like can make the difference between walking away with a deal or with empty hands.

Lesson of Success
— 52 —

Make contacts.

There are no small contacts. Just contacts. And any of those can open the door to huge deals. Take good care of your contacts.

CHAPTER 26

DEALING WITH DIFFICULT CUSTOMERS

Dealing with customers is as much an art as it is a science. No matter how well you treat customers and how good your customer service is, there will be difficult customers along the way.

Having great customer service, excellent communication skills, and outstanding procedures, among other qualities, allows you to reduce the frequency of dealing with difficult customers. No matter what you do, at some point you will have to deal with one.

The first step in dealing with an upset customer is determining what is making the customer upset. Over the years I've learned that some customers are frustrated and want somebody to express those feelings to. They need somebody to listen. They are not expecting anything, but just want to be heard. I have dealt with many customers like this. They get upset with the project managers or superintendents and start yelling or being rude, and it all boils down to them needing to be heard.

At Smarter Remodeling, the person above the project managers is a VP of Business Development, who is in charge of accepting the projects from the Sales Department and ensuring the project managers are executing projects on time and on budget. When a customer complains, he intervenes. He has a conversation with the customer and listens to their concerns and proposes solutions. The vast majority of the time, the solution is very simple. The customer is frustrated and just wants somebody to listen.

Oftentimes, the project managers deal with many projects at the same time, and occasionally communications are missed with the customers. When this happens the customer assumes they are not being taken care of and a cycle of frustration begins that culminates in yelling matches with the project manager or crews.

One recurring issue we had was when the project manager communicated an event to the customer but no updates were given on the issue to the owner. It created a sense of abandonment on the customer end that would accumulate and eventually erupt.

On the project manager's side, the explanation was that since there were no updates there was no reason to keep calling the customer. On the customer's side, there were no updates and they began feeling anxious. The situation could have easily been avoided with a phone call or a simple message stating, "No news yet but we will continue keeping you updated."

Always keep in mind that you know what's going on at the jobsite, but the customer doesn't. Establish systems that keep the customer aware of the progress made, or lack thereof. You will save much aggravation by doing this.

Lesson of Success

— 53 —

Keep your customers informed.

You have the power to change the outcome of the interaction regardless of how the other person approaches you. This is a very empowering thing to bear in mind.

Your behavior influences others' behavior, which is why it's crucial to employ certain skills to ensure a positive result.

It's your individual perception of "difficult" that defines the situation.

Lesson of Success

— 54 —

Don't label customers.

Don't allow employees to label a customer as "difficult." Encourage them to find what's making the customer appear as "difficult" and change the dynamics of the situation.

Allowing employees to label customers creates a pervasive culture of neglect.

Lesson of Success

— 55 —

Be proactive in providing great customer service.

You must be proactive to avoid uncomfortable situations with your customers. The vast majority of the time, customers don't want to have these conversations with you. But if you let small frustrations accumulate over time, you will have to deal with a problem of huge proportions, in which nobody knows where it originated or how it got so big.

It can start with the first call you didn't make to update your customer on the progress or lack thereof. The smallest details matter, so make the call. Keep them informed even when there are no changes to report. It will save you many aggravations.

SCAN ME FOR MORE CONTENT

CHAPTER 27

DELIGHTING CUSTOMERS

At Smarter Remodeling we didn't want to have one person who dealt with unhappy customers. Rather, we wanted unhappy customers to be as rare as white alligators, so we created our Concierge Service™.

The idea was to empower everybody in the company to remedy a situation that was making a customer unhappy or to do something that would make (or keep) the customer happy.

A few years ago, we were remodeling a laundry in a home with a family of five, including three small children. We had warned the customer several times of the inconveniences during the process, which included no access to the laundry machines for a few weeks.

We talked about this during the pre-construction meeting, again the week before we were to remove the machines, and finally the week of the event. Regardless of the notices, the homeowner, Ms. Ryan, found herself with piles of laundry so high she'd normally greet us by waving a small flag so we knew where she was.

Throughout the construction phase, Ms. Ryan was very nice to us. She was always smiling and telling us what a great job we were doing, and the crew was always in the mood to help her out moving furniture and household items so she could keep running her household smoothly. But they could see she was struggling with the laundry situation, and oftentimes they would see her digging through the piles trying to find an article of clothing.

Our *design concierge* had developed a close bond with Ms. Ryan and decided to take matters into her own hands. She hired a laundry service to come to the house and pick up all the laundry, get it washed, and return it all in the same day. When Ms. Ryan came home she walked by where the pile of laundry had been, not realizing it was gone. She walked into the living area and saw a set of flags waving behind neat piles of washed, folded, and categorized laundry.

Our crew was sitting there; smiling, waving the flags as Ms. Ryan used to greet us. She covered her mouth in amazement and after a few moments of holding back her tears, she ran to hug the crew.

There is no better feeling in the world than knowing you made a difference in somebody's life—even when it's something as small as doing laundry. We completed the job and she became a raving fan of ours.

The referrals we got from Ms. Ryan kept asking how she was happier with her construction experience than she was with her fabulous new laundry area. They told us that she was a raving fan and that they wanted the same experience. We always replied, "Our construction Concierge Service experiences are tailored to each customer and so no two are the same. Our crew is empowered to take every opportunity to make you a raving fan and we hope we can convert you into one, if you give us the opportunity to work on your project."

Most referrals do become customers, making us one of the most recommended contractors in the area.

Lesson of Success

— 56 —

Go beyond customer service

Good customer service is largely expected by most customers. In order to differentiate your business from the competition, you need to go beyond this and delight your customers by creating positive customer experiences. Your company culture should revolve around this and you should empower your entire team to create these experiences at every step of the process. From the initial call to project completion, everybody involved in the construction process should be trained and empowered to create unique, tailored experiences for each customer.

SCAN ME FOR MORE CONTENT

CHAPTER 28

---◀◆▶---

QUALITY GUIDELINES

You will encounter customers who are difficult to deal with because they obsess over extremely small details—the ones that can see a pinhole the size of a microbe ten feet from the wall and demand that you redo the entire wall.

Sometimes you can clearly establish what they should expect from the job. However, sometimes it doesn't work that way and the customer can become a thorn in your side.

To deal with customers like this, the best solution is to have written quality guidelines that specify exactly what they can expect from you. We must be as specific and as detailed as possible in our contracts, but it's impossible to detail the quality of installation of every single item in the project.

As an example, on the contract it might say:

110 lineal feet of WM 618 9/16 in. x 5-1/4 in Primed Finger-Jointed Base Molding, professionally installed in 8' increments.

This is specific enough for you as it describes how many lineal feet you are installing of what material. It even says you will be using 8' pieces (**Note for the super-contractors reading this**: I'm aware we can use longer pieces to avoid additional cuts ... and that by using the word professionally, I have left this open to interpretation ... this is just an example demonstrating picky people).

You know you will make the appropriate cuts to minimize the appearance of joints and that you will use wood putty, sanding them to make most, if not all, of the joints disappear. You know all this, but often customers don't. You know the product you will use to minimize the appearance of joints but your customer may have something else in mind because ... they saw it online! And of course, if it is online, then it must be true!

One of the thousands online keyboard warriors out there got paid to post an article about how one particular brand of wood putty was the best to the detriment of the one you like to use.

Now, after hours working on getting the baseboards looking great, you have finished the trim installation and are contemplating your masterpiece while thinking it's the best trim job you've ever done. You are even considering submitting photos of your work to "Baseboard Magazine" for consideration for the "Trim Installation of the Year" award.

You can already see yourself in the Ritz Carlton ballroom surrounded by hundreds of other trim carpenters from around the country, all of them holding their breath while the Master of Ceremonies opens the envelope to read the winner's name. You can see your competitors already bracing themselves to stand up to receive the award, but you already know you've won.

In fact, your name has already been called: "Mr. Jones." It takes you by surprise that the Master of Ceremonies didn't use your full name. Most people just call you Tim; but what the heck, you got the award. The guy is walking towards you, holding the award in his hands ... wait a minute ... that's not the Master of Ceremonies. It's the homeowner holding the empty container of wood putty he pulled from the dumpster—where you had put it a few minutes earlier!

Mr. Homeowner is walking towards you yelling, "Mr. Jones, Mr. Jones!" Now that he has brought you back to Earth and away from your award-winning dream, you focus on him and ask, "What's wrong, Mr. Homeowner?"

"What's going on?" He replies, looking at you as he was looking at Charles Manson. "You have ruined my job!"

And with that you went from living the dream to trying to survive the nightmare, the kind where you know you're dreaming but can't wake up from it.

After you get over all the yelling and accusations of being a trim murderer, you realize the customer is mistaken and you need to educate him.

Big mistake.

He pulls out his printed copy of "How to Properly Apply Wood Putty to Trim: The Only Way" by Keyboard Warrior. He won't be taken advantage of since he is now "woke." He knows the "right" product you should have used, how you should have properly installed the trim, and the proper way to make joints. He won't accept anything less than what Keyboard Warrior has said

You look at your trim job and squint your eyes trying to find a detail that would justify the verbal hemorrhage that was dumped on you, but you can't. The customer keeps pointing in the general direction of the molding. You start walking closer and closer and closer. Finally, you get on your knees, your face almost touching the area the homeowner is signaling at … and finally you see it, there is a faint line.

You continue to squint your eyes causing your retina to focus so you can see this line. You are trying to determine if there is an actual denomination for the size of this line: it's that tiny. It's definitely not 1/8", not even 1/16". You remember hearing of 1/32" and 1/64" but have never visualized them before. Now you know why, it's because you can't see them! You mentally add a note to your to-do list: Get an eye exam, ASAP.

You are trying to think how you can do better than that line. You still feel you deserve that "Trim Installation of the Year Award," but not if this homeowner has anything to say in the matter. Despite your best instincts and because you want to make this customer happy, you agree to remove the trim, install a new one, and use the product he wants.

The homeowner seems satisfied with that. He shakes your hand and says, "You are a good man, Mr. Jones."

It takes three days of removing and reinstalling the trim, spending ten hours a day on your knees trying to get the miracle wood putty to perform, and you have achieved the impossible.

Now, the joint trim line is, you dare to say, invisible. You have achieved perfection. The joint trim line is most certainly 1/64" or less. You are proud of it. You put a small tarp over your new masterpiece, holding it up with a couple pieces of remnant trim, and call Mr. Homeowner so he can rejoice in your mastery of trim science.

At this point you are certain this is a science and begin wondering if there is a Nobel Prize for this skill. You mentally start traveling to Sweden to accept the prize, but you get a nagging feeling and decide to wait on booking the airline tickets until Mr. Homeowner has given his approval.

Mr. Homeowner arrives and you remove the tarp slowly, like uncovering a long lost DaVinci and produce a musical sound with your mouth to add a dramatic effect to the presentation. Mr. Homeowner looks at it and remains quiet.

You know he is thinking how to apologize for having doubted your mad skills with a saw and is thinking on how to ask you to invite him to receive your Nobel Prize in Trim Installation. You are not a grudge-holding person and will consider his request. However, his silence is starting to make you nervous.

After a minute or so, which seemed like an entire day, he says "I knew that miracle putty was the right one to use."

"Sure, whatever," you think, feeling relieved to finally be done with this trim.

"Now, what are you planning to do about the line?" he asks.

Your frustrations begin to rise and you feel as though you're becoming the trim murderer and start coming up with ways to make this guy disappear.

You can't believe what you are hearing. The line is invisible. At least to the naked eye. Your phone cannot focus well enough to take a photo of the line. How in God's green earth can the guy even know where the line is? The guy has superpowers to see it. You calmly inform him that it's an extremely good joint and you cannot do any better: you have reached the limit of your humanity.

Mr. Homeowner looks completely disappointed and asks, "Well, can you get somebody else to do it any better?"

You are flabbergasted.

There is no human who can do this better. Maybe God can send someone from Heaven, but you haven't been to church in a while so you might not be in the best position to ask a favor from God.

You take courage, breathe in and say, "Mr. Homeowner. This is the very best this trim is going to get. Not me nor anybody else can do better."

He looks at you like you are an insignificant insect who dares to challenge him and decides to quash you right there. "Well, I guess we are at an impasse then because I'm not going to pay for substandard work."

Life is leaving your body. You spent four times the budgeted amount of time, used two times the amount of materials, missed one of your kids' baseball game because you couldn't determine the proper setting time of the miracle putty, and made your wife mad because you were late for dinner every night that week. To cap this inevitable corollary, Mr. Homeowner refuses to pay you.

You finally realize you will never make him happy. What now? You want to get paid, but he refuses. He argues your contract clearly states the trim will be professionally installed and if he can see a line then it's not installed professionally so *you* are in breach of contract.

Thankfully, this story is not yours. Why? Because you used a clear definition in your contract that refers the customer to a standardized third party to avoid the circumstances described above.

There is a widely used set of guidelines produced by the National Association of Home Builders (NAHB), a national trade group with a long trajectory that has a large presence across the country with the name of Residential Construction Performance Guidelines (these guidelines get updated from time to time, so always have the latest version).

At the printing date of this book, the most circulated version is the fifth edition. No matter which edition you use, properly make references to this in your contract so quality disputes will be decided by these guidelines, not by the subjective opinion of the customer or even by yours.

If Mr. Jones had included this in his contract, he could have told Mr. Homeowner that according to the guidelines referenced in the contract the openings and joints in trims and moldings should not be larger than 1/8"; and since the joint Mr. Homeowner was referring to was under this tolerance, no corrective action would be taken.

Imagine all the aggravation, time, and money Mr. Jones would have saved if this was incorporated into his contract. You don't have to write these guidelines, but you can buy booklets with these guidelines to give to your customers when the contract is signed.

Add the legal language to your contracts that incorporates these guidelines as part of your contract. You rarely will have to refer to them (as hopefully you already perform above these guidelines!), but when you encounter a difficult Mr. Homeowner, you'll be prepared.

Lesson of Success

— 57 —

Use standardized guidelines to evaluate work.

Once more, don't be like Mr. Jones. Use standardized guidelines to define the quality of work your customer should expect. This will avoid confrontations with customers but, most importantly, will save your lots of aggravation.

CHAPTER 29

RUDE OR VERBALLY ABUSIVE CUSTOMERS

Occasionally you will have a customer who, after being the winner-of-the-nicest-customer-award at your contract signing, becomes Mr. Hyde, going ballistic over minimal incidents. These customers can be verbally abusive by using expletives directed at you.

There are other customers who might not be rude to you, but they are monsters to your crew and your trade partners—this will eventually become a problem as you will lose employees or trade partners because of the customer's behavior. These employees/trades will refuse to show up to job sites, or (even worse) they will stop working with you, which will affect the other jobs you may have running.

Rarely, you can see the issues these customers produce before signing the contract. I'd give away my firstborn if there was a sure way to detect customers who will become impossible to deal with over time. Many authors will state there are telltale indicators where you can see the signs before you get into a contract with these people, but in all my years as a business owner, I have never seen the signs. When I see red flags, I just don't walk away. I run. I run like I have a pack of hungry lions chasing me.

Regrettably, though, most of the time the signs didn't manifest themselves until we were chest-deep into the project.

The best way to protect yourself from rude, obnoxious, and verbally abusive customers (I keep adding verbally to the abusive part because if you encounter any other kind of abuse by a customer, leave immediately and call the cops) is to have a clause in your contract dealing with this. This way, if a customer becomes rude or verbally abusive, you can have a conversation with them about the situation, refer them to the contract, and inform them how you intend to make use of the clause if the situation persists. Smarter Remodeling has a clause in our contract where notice is not needed, but we have found that sometimes having a conversation with the customer and pointing out the issues solves the situation.

Our clause reads like this:

> Owner acknowledges that Company has an expectation of, and is entitled to, perform work under this Agreement in a non-hostile environment free from belligerence, verbal abuse and profanities from Owner or anyone acting on Owner's behalf. In the event Owner or anyone acting on Owner's behalf is belligerent, verbally abusive, rude, hostile, threatening or otherwise uses profanities towards any of Company's employees, agents, representatives, or any subcontractor, laborer or materialman who performs any part of the Work under this Agreement, said conduct shall be considered a material breach of this Agreement and Company,

in its sole discretion, shall be entitled to terminate this Agreement without prior notice to Owner. Owner agrees that he or she has the ability to choose to refrain from being belligerent, hostile, verbally abusive or otherwise use profanities, and therefore does not need prior notice of Company's decision to terminate this Agreement in the event Owner is hostile, verbally abusive or otherwise uses profanities. In the event Company elects to terminate this Agreement under this Section, Owner shall remain liable to Company for all unpaid Work performed by or at the direction of Company.

Your legal counsel should review this and determine the best language to protect you and your people, from rude and verbally abusive customers. But always have some similar language in your contract.

Never engage in a verbal match with your customers. If they can't hold a mature conversation then it's not worth trying to engage. Enforce the rights established in your contract and walk away. There are plenty of fish in the sea, so don't waste your time with verbal piranhas.

By the way, it doesn't matter if the issue that precipitated the verbal abuse was yours or your crews' fault. There are mature ways to handle issues, and the use of profane and abusive language is not one of them.

Lesson of Success
— 58 —

Do not accept abusive behavior.

Do not engage in verbal matches with your customers. You cannot win by showing your wits and verbal skills. It's always better to let the customers calm down and then establish a conversation conducive to resolving the issue. If a conversation cannot happen due to the customer's unwillingness to, then why are you staying at that job? Terminate your contract using a clause like the one above and move on.

CHAPTER 30

SURVIVING TOUGH ECONOMIC TIMES

One time, I went to a business conference in Las Vegas. It was mid-morning and we had finished attending a long educational seminar (which was extremely boring, but the breakfast included made up for it). We were trying to determine what seminar we would attend before lunch. We saw a business guru seminar which seemed to be filling up fast. I had my doubts but, once again, the buffet table made the decision for us.

We headed in, and after grabbing some drinks and more food, the lecture started. The guru started talking about the recession we were just leaving and how the tough economic times forced many companies to close. He went into statistics and numbers about the damage the recession caused to the construction industry and the country's economy in general.

I wasn't interested in the topic, and faded in and out of focus, until I heard the speaker say: "The best way to determine the resilience of a company is to go through some tough economic times."

Did I hear the guru correctly? Was he suggesting that companies who go through tough economic times and survive, then they are resilient; and if not, they aren't? I wasn't sure if I heard him correctly or caught the phrase out of context.

The phrase would continue to haunt me. It shook me enough to prepare my company to withstand unexpected turbulent times so we could survive—and even thrive—during economic downturns or markets with high volatility.

It is imperative you prepare your business for a rainy day or season. It's not a matter of *if,* just a matter of *when* these times will come. And you need to be prepared.

Of course, if these times are precipitated by a global pandemic with a reach beyond anybody's wildest imagination, then you can give yourself a pass if you fail or do what Smarter Remodeling did: find a way to turn it into an opportunity to grow and expand.

Keep reading to see some of the things we did to thrive during the pandemic and how you can apply some of them to your business

---

Lesson of Success

— 59 —

Be prepared for hard times.

Be prepared for difficult seasons. Be ready not only from a financial standpoint, but creatively pivot so you can survive an economic downturn—and even thrive.

We spend so much time lamenting over things we have no control over. Why don't we spend time coming up with ideas to change the things we can control? Like always, pick something small and get started. Don't just think about it anymore. Do it.

CHAPTER 31

---◼◆◼---

ACTIONS FOR DIFFICULT TIMES

I repeatedly hear people say that failure makes you stronger. While I believe learning from your mistakes makes you wiser, I'm not sure that failures by themselves make you stronger. I believe getting up and pushing forward every time you fall shows the true strength of your character. We are not defined by our failures but by the strength shown in recognizing our mistakes *and* changing.

Learning this almost cost me everything.

After years of working hard I had finally been able to create a somewhat profitable business. I hadn't yet learned the many other lessons I would learn over time which would eventually make me very successful, but I was enjoying some moderate success at the time.

For instance, I hadn't learned the importance of tracking everything in the company so I could rely on numbers to make every decision rather than instinct. Part of this was the problem that I hadn't started tracking every job-related cost so I could know which jobs were profitable and which ones were not. I had relied on being profitable overall, which it was a bad habit I would come to regret.

*****

Several years ago, I was at the point where I wanted to grow and be seen as a serious competitor by other contractors in the area. I had realized the importance of branding (although I hadn't yet come to the realization that we needed key differentiators from our competitors to stand out in a crowded field), and my company name was getting traction in the right circles.

Within my circles, a certain man knew of a project to remodel a local shopping mall. He had knowledge of the contractors bidding on the work and felt Smarter Remodeling could offer a competitive price. I made the huge mistake of competing on price and, as a reward for my stupidity, I got the job.

The shopping mall remodel consumed almost every resource and every free minute I had for months. I couldn't quit because my company name was at stake, and I wasn't willing to ruin our reputation. I tried discussing with the owners to renegotiate the contract so Smarter Remodeling could break even but they wouldn't hear it.

I was falling apart. I missed my kids' games and school functions. I was never home and when I was, I was too tired to even interact with my family. I was always in a foul mood as nothing was going my way. I was struggling to pay my bills and could see the company I gave my sweat and blood to build was crumbling.

Most importantly, I could see the writing on the wall of a marriage gasping for air to survive. I was in a vicious circle of depression and self-pity that only drove me deeper into nothingness.

One day, while I was sitting at my desk at home, my wife, Agata, brought me a thermos of coffee and asked, "Haven't you had enough?"

I looked at her a bit confused. "Enough what?"

She sighed and said, "Martyrdom."

Without paying attention to my taken-aback look she continued, "We all know things are hard and you're struggling. People and businesses go through struggles all the time. But you won't fix anything by sitting here wallowing. Enough is enough. Get out of here. Go out there and be brilliant. Be you. And find a way out of this mess."

She slammed my truck keys on the desk and marched out of the room.

I didn't have time to complain. Agata didn't give me the opportunity to keep playing "poor me."

I picked up my keys, the tumbler with the coffee, and headed out. I was mad. I sat in my truck and felt so enraged that I couldn't start it. I was ticked off. How dare Agata talk to me like that! After all, I had built everything we had gotten so far. It had been all me. I did all the work. I sweated. I bled. I. I. I. I kept thinking about how I had done everything—like she didn't play a huge role in that—and how I deserved compassion for what I was going through.

Suddenly a light penetrated my self-pitying foggy brain and hit the only neuron that wasn't pushing me to complain: I realized I had been far too self-centered in thinking how the situation had affected me—how everything revolved around me. Never for a second had I considered how the situation affected my wife, my kids, and my employees. I had been submerged in a sea of commiseration. I hadn't been paying attention to the people around me who relied on me for guidance, leadership, and emotional support.

The realization brought tears to my eyes and my gross mishandling of this situation paralyzed me. Thankfully, more and more of my neurons were waking up from the funk and started pushing me to take action.

I opened the door of the truck and headed back to the house. As I opened the front door, Agata stood in the living room. She was surprised to see me. I walked towards her without talking and she dropped a rag she had in her hands. I got to her, looked into her eyes. "Thank you," I said, and kissed her. Then I hugged her for a while.

"I know what I need to do now," I said. I headed back to my truck.

I headed to the mall. It had been a job that brought me to my knees but when I got there, I felt like I was in for an epic battle for survival, a modern-day version of the biblical story of David and Goliath.

I walked around the mall for hours trying to determine how to proceed next. It was to know what my next move should be as I was locked into an unprofitable contract with an owner who refused to renegotiate.

While walking through the mall I noticed several stores that were closed with announcements of new brands coming soon. I had my divine inspiration.

I remembered we had included a last-minute clause to the contract that dealt with increase in prices for materials—and after reading what the lawyer had sent, I added a few lines dealing with unexpected increases in labor prices. I ran to the exit, got to my truck, and headed to the office to find the contract (we hadn't reached the cloud age yet!).

I ruffled through folders and found the contract. I read through the contract and found my savior clause. I felt such a relief finding the clause, but at the same time I didn't want to use a technicality to get out of my contract. I decided to make an appointment with the owner to discuss how to proceed.

He wasn't too happy when I presented the documents that would allow me to increase the price of the job or to walk away. I stated I didn't want to walk away but also didn't want to continue working on the job while losing money. However, I had thought of something that would benefit both of us and allow us to keep moving forward with the project.

I understood he had entered into the contract with a predetermined budget. Regrettably, outside factors had conspired to make us change our prices to turn a profit; but I'd be willing to complete the job at the price agreed-upon originally *if* we were granted the second phase of the project at a competitive rate and if they would endorse us as the preferred contractor of the mall so we could have additional advantages when competing for the new stores' buildouts. "After all," I concluded, "You already have seen the quality of work we produce and the way we have been dealing with issues we've encountered."

The mall owner looked at me with a serious face, leaned forward, and shook my hand. "Deal," he said.

Now I felt like crying, but I didn't.

He continued, "Actually I believe it takes a good two months to properly vet any contractor and no contractor can work in the mall—starting today—without being vetted and approved by us. Since you are already vetted and approved, that should give you a nice head start over other companies."

I was floored. The man who refused for months to sit down to renegotiate a fair price was now giving me the keys to the kingdom. I smiled and thanked him.

"Why are you doing this?" I asked.

He looked at me now like I was his son. "Because you stuck to your word, acted with integrity, and even when you found a way to legally force us to pay you more, you came back with a solution that would work for everybody. That's the kind of people I like to be associated with."

With that he stood up, excused himself, and directed his people to arrange all the paperwork to put into motion what we had just discussed.

In the following years we completed several projects for the mall and for the new tenants that allowed us to solidify our reputation in the commercial arena.

You will always face problems; the important thing is how you deal with them.

Lesson of Success

— 60 —

Don't ever put work over family and friends.

A job is not worth losing family and friends. Your work can put you in a grave but it won't shed a single tear for you. Your family and friends will. Family should always be first and friends a close second.

Lesson of Success
— 61 —

Lose the self-pity.

Don't drown yourself in self-pity because things are not going your way. Take action to change and to produce change. Things won't get better by themselves. You need to take an active role in making things happen for you.

Lesson of Success
— 62 —

Make everyone win.

The best deals are the ones where everyone wins. If you only make deals when only you win, you will be dealing with disgruntled parties long after the signing. You'll get much further ahead dealing fairly with others rather than taking advantage of others.

Lesson of Success
— 63 —

Price increases and contract clauses.

There will be times when it's necessary to increase prices due to unforeseen circumstances in a job and it's vital that there are clauses in your contracts to make allowances for this otherwise you will not be able to do so. Check with your legal counsel to see if you can use something similar to the following clauses:

Material Price Escalation Clause: If, during the performance of this contract, the price of materials increases over 5%, though not the fault of the Company, the price shall be equitable adjusted by an amount reasonably necessary to cover any such significant price increases.

Reliance on Subcontractors Clause: Company relies upon subcontractors to perform the Work for the stated Contract Price. In the event that a subcontractor cannot perform any portion of the Work for the price quoted by the subcontractor, the Company, in its sole discretion, may cancel this Agreement or may increase the Contract Price if the cost to complete the Work quoted by the subcontractor exceeds 25% percent of the quote provided by the subcontractor and relied upon by the Company.

SCAN ME FOR MORE CONTENT

CHAPTER 32

OWNER COMPENSATION
PAYING YOURSELF

As an owner, you are entitled to additional compensation for the risks you are taking to invest in your business. This compensation is above and beyond the salary you normally receive for the time you put in your role within your company.

You get a salary for your role as a carpenter, or foreman, or administrator or general manager. Someone has to do that work for running the company and if it's you, then you should be paid for it. In addition, you have put up a fixed amount of money and probably personally guaranteed some loans for the business, such as to buy a building; to purchase trucks and equipment; to hire employees; and to advertise. To keep the company running in lean times, you might have to loan some money to the company. Because of this, there should be extra compensation for you.

The computation is simple. If you have invested $200,000 in the company and with an average rate of return at 4%, that comes out to $8,000 a year extra compensation for you. Consider it as though you had invested that money in something else. Say that you are on the hook for another $300,000 in loans. Then your exposure is $500,000 and at 4%, that comes to $20,000 in extra compensation.

You don't have to use 4%, it could be 8% or 10% or even 15%. Use a percentage that is justifiable as an adequate return for your investment in the company. If you normally make 15% in the stock market, then 15% is good. The percentage you use must be fair and something the company can afford to pay without hurting its normal cash flow. You should take this extra compensation for your funds and credit power. (In reality, we all know, it goes toward buying those 55-gallon drums of TUMS you hide in the warehouse to help with the acid reflux produced from years of not making a profit!)

While some people believe owner compensation should be included in the salary you get as an owner, I believe you should receive compensation that's commensurate with what you do for the company. This allows you to use the same salary to hire somebody else if you decide to stop working and start enjoying an early retirement. But on top of that, you should receive a return on your investment that is above and beyond the profits your company produces.

You may want to talk to your accountant and see if this compensation could be considered as an interest expense and could be made deductible.

*****

Owner compensation should be included as a percentage in your overhead expenses so that it doesn't affect your net profits.

This will provide a nice little chunk of money for you, and won't increase your overhead expenses significantly.

I have been following this system for years. I also know other construction business owners who don't, as they argue that it is the same as getting profits. I don't disagree. But, for myself, this is a way to get a return on the investment I made in the company in addition to the profits the company generates.

Think about it and consult with your accountant and decide if this could work for you.

Lesson of Success

— 64 —

Pay yourself.

I see many business owners making the mistake of working in their business and never collecting a paycheck. When asked why they respond they get paid with profits. If you are not paying yourself for the work you perform in your business then you might be masking an issue. When you might want to step aside your "profits" would now have to cover the salary of the person that would replace you. Are the profits enough to cover that salary and leave some for the owner's compensation? If the answer is yes then you will be fine. If the answer is no, you need to reevaluate your margins to increase your profits and consider start getting yourself paid.

CHAPTER 33

WORKING WITH A SIGNIFICANT OTHER

I resisted working with my wife, Agata, for years. I was convinced the stress of working and living together would destroy our marriage.

As a business leader, I failed to recognize her talents and the huge value she would bring into the company.

As a husband, I failed to see beyond the mundane and recognize her as a true partner.

Luckily, one day many years ago, I finally realized that not only we could use her, we desperately needed her. Her talents went beyond her charming personality and trustworthy persona; her true talent resided in her innate ability to bring to life our customers' vision combined with a sincere desire to make a positive impact in their lives.

It was a powerful combination that would change the way our company approached design and customer service. In time, we created the position of Design Concierge™ and the concept of the Concierge Service™ based on the way she approached every job and every client.

Agata is a true advocate for our customers and will battle suppliers, subcontractors, city officials, and others (which regrettably sometimes includes me!) to provide our customers with unique construction experiences and, in doing that, she forges bonds that transcend the typical company-customer relationship, making customers a part of our family.

It was a sight to see when Agata would face subcontractors or plan reviewers who weren't performing to her expectations. She was not intimidated. She reminded me of Joan of Arc fighting against the English. One tiny, little woman bringing entire armies to their knees. And although it was impressive to see her fight for our customers, it was nothing compared to her charm: people didn't stand a chance.

She remains, as of today, a fundamental piece of our company's culture and its most fervent defender.

*****

Working with your spouse may present challenges, but, when handled appropriately, it will be the most rewarding experience you'll ever have.

The key to successfully working with a spouse is to establish early on a balance between home and work life. You can't survive this relationship if you cannot separate the two.

Some dynamics at home might need to change or, at the least, adapt when both parents are working in the business. Agata and I struggled with this initially as household chores went undone for a while as we were both tired at the end of the day. We finally reached an agreement to split chores, while assigning some to the kids, and peace returned to our humble abode.

On the positive side, our relationship has grown on many levels as she now has a deeper understanding of the troubles and tribulations I went through building the company. She teaches me to see through situations from another's perspective, and I can see through her eyes how we can still improve things to keep building a better company for ourselves, our employees, and our customers.

Many couples struggle to find time to spend together and working together affords you that opportunity. Sometimes it's driving together, and sometimes it's having lunch together. Whatever it is, take advantage of the opportunity and make it count.

If you are thinking of bringing your spouse to work with you, I encourage you to take the plunge and do it. Set ground rules to separate work from home and start working to find the right place for your spouse. Don't fall into the trap of putting him or her to work doing administrative tasks just because you don't want to deal with it. Find the right position where he or she can thrive and produce the most impact in the company.

Your spouse is a very valuable asset. In my case, I can rely on and trust her. Her opinion won't ever be filtered, so, like it or not, I hear all the truth and nothing but the truth.

Remember, you are not just husband and wife but true business partners, so make sure you treat your spouse with the same level of respect and afford her the same courtesy you would afford any other employee or partner. If you don't, you'll pay the price at both work and home.

Lesson of Success

— 65 —

Work with your spouse.

Take advantage of the opportunity you have to work with your spouse to spend some quality time together. Put your phones down at lunch unless it's to plan something together and try to engage in a conversation during the times you commute together. This would make for a much more enjoyable time at work.

SCAN ME FOR MORE CONTENT

CHAPTER 34

LEAN, FATTY, OR MARBLED STRUCTURE

I grew up in a country that produces some of the best beef in the world, so I'm very familiar with the concept of lean, fatty, and marbled. For the vast majority of beef connoisseurs, marbled cuts are the best of the best and varieties like Wagyu beef have made an entire new category over it.

Out of my fondness for this food of the gods, I choose a corporate style that resembles the best of both business practices: lean and fat (recall phrases like "we have to cut the fat" when companies are trying to cut costs). I call my style: marbled.

What is a marbled business? I don't want to run a company by counting pennies in all the processes we use. At the same time, I didn't want to waste finite resources. To begin, I started by adopting a methodology of hiring people who are able to generate revenues (resources) and not only people who will administer the revenues currently being generated.

This has allowed me to keep hiring talented people who help me keep growing the company and at the same time have allowed me to look for new business opportunities.

Simply put:

On the fat methodology you hire people no matter what they generate to help you manage the chaos inside your organization.

And, on the lean methodology you only hire people when it's strictly necessary—and since you're counting pennies, then you're hiring whoever fits your budget and not who is the right person for the position.

With my marbled methodology we hire by focusing on who is the right fit to the company and who gives us the ability to increase our capacity to take on more work or generate more revenues. This kind of talent normally comes at a higher price and we are willing to pay it. By doing so, it allows us to generate growth and perpetuates our corporate culture of profitability as a means to service all our stakeholders.

We also mainly hire in the management positions or in positions that can become management. For the most part we don't hire technicians. It is more important to generate management positions so our employees can become leaders and help us improve our processes and systems, which in turn help many technicians to become their own bosses and run successful businesses.

It also gives us a lean organization that responds quickly to unpredictable market changes. Our management is trained to use *all* the resources at their disposal to create company revenues that can be used to help more and more people. By establishing this structure, we get the best of lean and fat, and get a beautiful, efficiently run, "marbled" company.

I guess you can say I'm obsessed with beef. Well, I am.

MARBLED + COVID-19 = THRIVING

Nobody expected COVID-19 to reach the dramatic levels it did by the summer of 2020.

In February 2020, I was flying to Hawaii for my daughter's wedding and I was seeing people at the airport wearing masks. Not many, but a few. I wasn't surprised, as I had just traveled across Asia, where many people were wearing masks inside and outside of the airports.

We started our *looong* trip from Jacksonville to Hawaii with my eldest daughter, her husband, and two kids. In addition, I was traveling with my wife, youngest daughter, and son. We had negotiated for quite some time with my daughter Karen about a big, local wedding or a smaller wedding in an exotic destination.

When looking at the costs of having a local wedding I was floored by the sheer amount of details it involved (and the cost!), so when I was presented with the option of a smaller wedding for only immediate family, I was ecstatic.

That happiness only lasted until I saw the costs of flying, lodging, ceremony, clothing, feeding, and entertaining the immediate family in an exotic destination, but, heck, luckily we had built a company that allowed us to afford all this, so we did it.

It was truly moving to see my second daughter get married. I performed the wedding ceremony of my first daughter, by the power vested in me by the state of Florida, and it was a unique experience; but since I performed the ceremony there were many details I missed (though I saw them later on the video). But standing a few feet away from Karen with my feet deep into the warm sands of Hawaii was fantastic.

The Hawaiian views are hard to describe. I always say you have to *feel* them not just *see* them and I stand firm by this. Beauty like this cannot be accurately described with words.

Neither could I describe how gorgeous my daughter looked in her floating white wedding dress. When we got out of the limo to start the walk to the beach where the ceremony would be performed, I had to hold my tears back: seeing her reminded me how spectacular my daughter Alexa had looked on her wedding day ... and how now my tax deductions were getting cut again. Just kidding, I'm not that heartless ... but I could still claim her for that year!

Seriously, it is a bittersweet feeling, being proud of my daughters going into the world and starting families of their own and the sadness of seeing our household get smaller. Alexa had gotten married the year before, and now Karen. Soon my daughter Kiara would leave for college and Agata and I would be left with just Andrew, our special needs son, who deserves his own chapter to go over all the lessons he's taught us.

We had a blast on that trip. The wedding was absolutely gorgeous. Everybody looked fantastic, even me, with my slightly overweight butt (this is my book and I can take as many literary allowances as I want!).

We toured the island, created amazing family memories, and then Karen left for her honeymoon in Punta Cana and we headed back to Jacksonville. Unbeknownst to us we were about to face some trying challenges and go through a pandemic that would change millions of lives forever and would affect us in more ways than we could imagine.

We got home at the end of February, and COVID-19 still was something we thought we would never go through, or was one of those "viruses" on the other side of the world the media overexposed to attract viewers but never made it to the mainstream here in America.

Two weeks later, in the middle of March 2020, the World Health Organization declared COVID-19 a global pandemic and we went into massive shutdowns prompting business closures across the nation. It was nothing we had ever seen before.

Within weeks of coming back from an idyllic trip to Hawaii we were forced to stay home, unable to leave unless it was absolutely necessary to get food or to receive medical attention. It was surreal. We went from being on top of the world to being crushed by the weight of it.

The first few days of the quarantine were great! I had all the time in the world to catch up with my favorite shows, lay on the couch without a worry, play some games with my kids, and watch more TV.

Since everything was shut down there were no phones ringing, nor a pile of emails to respond to. Nothing. It was the best staycation ever … for the first 3 days.

It didn't take long until I realized that this was going to be longer than a week or two. I started getting concerned with how fast the pandemic was growing. We weren't overly concerned in Jacksonville (as we didn't have a lot of cases), but things started to look dire.

I started going to the supermarkets very early in the morning and was amazed by people queuing to get in. When the doors opened it became a Black Friday revival with people running towards the aisles with toilet paper. I never understood why people needed that much toilet paper but I guessed there might have been a psychological comfort to have it. I remember grabbing a pack like we always did and when I got to the cash register the lady told me, "You are allowed to get 48 rolls of toilet paper in case you want to get more."

I'm a bit sensitive about my slightly overweight butt so I looked over my shoulder and replied, "But I still only have one no matter how big it might look."

The lady looked at me confused. I tried to explain the joke but she got more confused so I said, "I'll go get another pack."

By the time I got back to the aisle there was not a roll to be found. Nothing left. Even the paper towels were gone. I assumed there might be some rough butts out there, as no matter how big mine gets, it still has sensitive skin and I would not use a paper towel unless it was an absolute emergency.

I went back to the register and found my cart moved from where I left it—and the pack of toilet paper was gone. It wasn't that I had needed it, but really? Stealing toilet paper from another person's cart? Was this pandemic already causing people to lose sight of the rules of common civility?

If that was the case, I shouldn't be concerned about toilet paper. I should start buying plywood to board my windows and bullets to defend my property. If only I had bought plywood at those prices and held it for a while: I could have quadrupled my money as the price of construction supplies started skyrocketing!

I returned home with groceries but no toilet paper (or plywood and bullets!), but was determined to protect the right of my butt to be wiped with soft tissue and made plans to return to the store the following morning just to get toilet paper. I had become a toilet paper warrior. And I conquered. The next morning, I walked out of the store with four packs, not two of toilet paper. Now I was the proud owner of 96 rolls of toilet paper. I gamed the system and protected my divine right to have a bountiful provision of toilet paper.

My delirious reasoning lasted until I got to my car and realized I couldn't fit all 96 mega rolls of toilet paper in their packs. I had to break the packs apart to use every available space in the car. I made it work but it also made me realize that I had let myself be consumed by panic buying that made no sense, and that I had only contributed to the overexertion of an already struggling supply chain.

I atoned for my sin by distributing the toilet paper among my employees who had issues getting rolls, because people like me were hoarding them.

The panic buying experience got me thinking about my customers and how they were also having issues getting supplies: toilet paper, water, disinfectants, etc., and how my crews were going through the same problems.

I started reaching out to my contacts across the country and internationally to source alcohol gel, masks, gloves, and UVC disinfection lights. I was able to locate what I needed and arranged air freight to get it to Jacksonville, ASAP.

Smarter Remodeling was one of the first companies to have masks available for our employees, as well as alcohol gel to disinfect hands at all job sites. Since people were afraid of letting strangers in their homes, we started promoting virtual consultations. We provided our sales staff with masks, shoe covers, gloves, and disinfecting gels both for themselves and to leave with customers who wanted in-person visits. We provided our crews with disinfecting UVC lights to disinfect tools and work spaces.

Once we received our shipment of masks, we provided them to all our trades, employees, and customers and then donated the rest to first responders who were struggling to get protective equipment. We imported thermometers and donated hundreds to individuals and schools.

We were considered essential personnel and were able to go to job sites and continue working on open projects. However, our workload had dropped significantly. With extra time in our hands, we decided it was time to get all our systems and procedures fully documented and started seriously working on Smarter Remodeling 360 Solutions.

At the same time, we decided to get serious about managing our data and began developing our own software. The plan was that this software would become a powerful construction management tool at the center of our platform.

We didn't commute daily to the office, so we had more time to sit down and write and write some more. Between March and June, we put more systems, procedures, plans, and dreams in writing than in the previous five years combined. All that writing is the basis for SR360 Solutions.

A few months in, I started to get worried, as we hadn't sold anything in four months: March, April, May, and June. Suddenly, in July 2020 we started closing contracts again. We closed, even with periods of very low revenues, the year with gains over 2019. We closed 2021 as the best year in our company's history, ahead of the launching of our new venture, Smarter Remodeling 360 Solutions, in February 2022.

The pandemic didn't change us. It accentuated our resourcefulness and creativity and allowed us to come up with new ideas to entice potential customers to keep working on their remodeling projects.

We didn't lay off a single employee during the pandemic; rather, we grew our team—which explains why we had our best year yet, as every hire not only helped manage existing business but also brought in new business. (Like I said earlier, this is why we only hire management and not trades. We train the trades so they can become profitable businesses and reliable partners for us!).

It doesn't matter if you are confronted by tough economic times or a global pandemic. What really matters is being flexible enough to respond to events and having a team of people who totally identify with your company culture. Otherwise, you will find yourself trying to guide a cart being pulled in every direction but moving nowhere.

Take your time to hire, but when you do, make sure your new employee can learn from you and you from him/her. Make sure that their talent resides in being open-minded and ready to take on challenges. Many employers focus on only the people who have the right experience for the job. Smarter Remodeling doesn't follow that pattern: we mostly hire people without construction experience but with the right personality and combination of talents so they can thrive within our company culture. We can teach construction, but we can't provide the right personality or innate talents.

Never be scared of the unknown. Be scared you are taking too much time to make your move. Go ahead. Make waves. I don't even know you but I already believe you can do it. After all, I did. And nobody believed in me.

Lesson of Success

— 66 —

Believe in yourself.

You have, inside you, all the strength you need to overcome any challenge. You just need to reach in, find it, and put it to work.

Lesson of Success

— 67 —

Train your employees.

Before hiring you should have ready the training material for the new employee. Hiring without this is just wasting hard earned money. Many contractors hire and assume the new hire will be trained by existing employees even when there is no formal training in place.

If your current employees are not properly trained how do you expect the new ones to be?

Failing to make plans is planning to fail.

CHAPTER 35

———•———

WHAT SMARTER REMODELING 360 SOLUTIONS DOES

SR360 Solutions can do many things for its members but the most relevant question is: *why?*

To us, each member is not another unit of a global business. They are part of our family. We commit ourselves to their success the same way we commit to our own success. Every day. Each member can expect to interact with us, our team, and their Power Group often so we can continue to help build their success and to hold them accountable for their success.

In return, they can expect to help shape the future of the business with suggestions and ideas for growing and improving.

Each member is a key piece of our organization. Each member can expect to be treated as the most important member, because for us they are. They will be from across the country, representing their values and building their brand, while using our systems and putting into practice what they have learned through those systems.

Our fees may seem more than other software technologies available, but there is a very important difference.

We are not just a software company. Our platform has the tools to empower your business to become more efficient and profitable. We created our software to manage your data, but also to give you the tools to produce it through ongoing training in key areas, such as marketing, sales, and business.

SR360 Solutions members can expect to see new software features, new systems, and new procedures being constantly developed in service of increased profitability and efficiency.

We are not a company who will sell you a membership, give you a book, and wish you good luck.

We are a rare business that's completely committed to your success. From the initial training to help set up your software, to specific one-on-one coaching to help develop sales strategies that turn into closing bigger deals, to helping recruit to run projects, we are there for you every step of the way. You are in business for yourself, but not by yourself.

We are in this together.

Visit us at www.sr360solutions.com

SCAN ME FOR MORE CONTENT

EPILOGUE

---◆---

THE FUTURE OF SR 360 SOLUTIONS

Year 2032. Sitting behind my desk, I'm looking at a series of screens showing, in real time, every new customer, proposal, project, and expense happening throughout the entire software that now has over 100,000 users; this always gives me a sense of pride and accomplishment.

Our platform has become the gold standard in the construction industry. We chose to focus on offering a way to manage the data and also mastering the skills to meaningfully produce it, through the development of marketing, sales, and business systems and procedures conducive to improving the overall efficiency and profitability of the business. While others were only interested in selling software, we concentrated on ensuring our users were successful by teaching them the skills they needed to fully use and take advantage of every aspect of our software platform.

When we first launched ten years ago, we dreamed of moderate success in helping other construction companies achieve consistent and predictable profitability, but we never imagined we would, before our tenth anniversary, be looking at an organization with revenues in the hundreds of millions.

Looking back, our success was built in our core belief that helping others succeed was the surest path to our own success. We built this organization around this core value and we made sure every new hire fit into this culture.

It is a culture that revolves around relentlessly pursuing profits in the service of every one of the company's stakeholders: employees, customers, vendors, shareholders, and the community at large. We hired, not based on experience, but on the commitment, willingness, and aptitude of the new hire to put his/her talents into the service of helping others achieve success. By doing that, we transformed a self-serving industry into a beacon of hope for the rest of corporate America.

Looking back to 2022 when we launched SR360 Solutions, I realize we were as right then as we are today: focusing on ensuring our users' profits and not on our own. In doing that we showed our true commitment to the success of our brand—and of our users—thereby paving the way for our own success.

We formed true partnerships with our users and together we have reshaped an industry plagued by bad actors, scam artists, and unprofessional players, into one where we can be *proud* of being a construction business owner.

We have done this in an industry where, after decades of lackluster recruitment, kids are graduating high school and pursuing construction-related degrees, because we have shown them we can change lives by providing a secure home, an accessible home, a livable home, or just by adding spaces conducive to more family gatherings. We have shown that we have the power to enhance the quality of life for our seniors and newly formed families.

We have shown the country our industry can truly make a difference and we did that by instilling our core values in each one of our users and by allowing them to teach us their values. We asked them to share their dreams and their hopes; we made them our own because that's what families do. They dream together. They hope together. They achieve together.

At Smarter Remodeling 360 Solutions it was never about getting to the top alone; rather, it was about getting there surrounded by the people we helped along the way and the people who helped us get there. We are strong by ourselves but we are invincible together.

My only regret is not having started this journey earlier, because we could have helped so many more entrepreneurs to flourish and become success stories.

For themselves.

For their families.

For their communities.

For this country.

THANK YOU!!

LESSONS OF SUCCESS

1. Customer experiences and kindness.

Be kind like my son Andrew, who did this with the not-so-nice salesman from the Porsche dealership.

Strive to provide exceptional customer experiences, just as the salesmen from the BMW dealership. They went above and beyond to make me feel appreciated during and after the purchase. Therefore, I have remained a loyal customer to them over the years and I speak about them positively all the time. I also frequently speak about the other salesman—but not in a positive light.

Go beyond customer service and embrace creating customer experiences. experiences

2. Live your own American Dream.

No matter what you think the American Dream is, make plans and take actions so you get to live your own version of it.

The beauty of living in the land of the free is we get the opportunities to make our dreams a reality. Don't let them pass. Summon the courage, take a leap of faith, and build your American Dream.

3. Be careful who you lend your ear to.

In the story above, the "little devil" gave bad advice, and since it was the easiest thing to do, I followed it.

Oftentimes, the path of less resistance is not the right one to follow.

Surround yourself with people you can learn from and whose sage advice you can follow. Beware of false "gurus" or "little devils," and remember, there are no shortcuts for doing the right thing.

4. Dream big.

I encourage you to dream. But dream with an abundance of details. Break the details of your dreams down to a list that can be used to track your progress towards achieving your dream.

But most importantly remember to take action. One small step and then another. Dreams with actions become reality.

5. Have all the facts.

Have all the correct facts if you are going to create a story. But, it's more important not to lie to get out of your obligations.

Work smarter so as not to come up with stories.

I have built a business that allows me to work whenever I want, from wherever I wish, without the company suffering in my absence. Create a business that is self-sufficient without your expertise. Work hard towards your own reality, molding your future the way you want it.

6. Honest work pays off.

I took Grandma's work ethic very seriously and worked very hard to achieve my goals. One thing I learned in all my ventures is that to be a great leader you should have worn all the hats in your company. Nothing will replace firsthand knowledge of every position, and this knowledge will allow you to make better decisions when creating systems and procedures for that position.

7. Read. Never stop learning.

It is the best and cheapest education you can give yourself.

Open your eyes and ears and learn from those around you. You could be in the presence of greatness, but if you don't open your eyes to see and your ears to listen, you will never know.

8. Listen to your elders.

The wisdom they've learned throughout the years can save you a lot of aggravation. You don't have to take their advice word for word, but listen carefully and make your own conclusions. It will save you a lot of time and money not making the same mistakes they made.

I often hear you have to make mistakes to learn from them, but I don't believe that. I believe you will make a lot of new mistakes that you can learn from; there is no need to repeat the ones others have made before you.

9. Focus on improvement.

No matter what area you work in, focus on what you need to do and find ways to do it better. Think outside the box; better yet, don't even think there is a box you need to fit into.

There weren't any snips made for kids' hands. They were all big and made of metal, but I found a way to make them fit for my hands. Long after I left the farm other kids and some of the women in the crew were using the modified version of the snips. I had made a difference in the lives of others without realizing it at the time (and I was still a child). There is no age minimum or maximum for innovation.

You can be brilliant regardless of your age. Your mind is timeless.

10. Never let others decide when you've had enough.

Even when it's out of love. Uncle Francisco allowed me to sleep in on the third morning of work because he loved me and wanted me to rest. However, by doing so he was depriving me of the opportunity of proving I could rise to the challenge. I had no intentions of making a career out of harvesting grapes and he knew this. He always said I needed to focus in school, but I wanted to prove I could do the physical labor and I'm glad I did: these lessons will stay with me for the rest of my life.

Sometimes people—with the best of intentions— will steer you in the wrong direction. Stick to your convictions and remember that no one but you can tell you how far to go.

11. Hard work pays off.

After working in the vineyard for a week I had earned enough money for sneakers, was able to give some money to Mom, and also had some extra to go out to the movies with friends. It was all gone in a short time, as I hadn't learned about savings or creating capital, but those lessons would come at a later time in my life.

12. Always listen. Learn the facts before judging.

During my time at the farm, I learned about the hard life of the migrant workers and their families and learned about discrimination. These lessons would affect my thoughts about immigration and influenced my preconceptions about immigrants.

Immigrants often get misjudged because of race, nationality, and economic status. We can be quick to forget that they are the ones who are actively seeking a better life for themselves and are willing to put in the hard (often manual) labor with long hours to ensure the improvement of their lives.

I saw how other people treated the migrants and accused them of stealing their jobs but quickly quieted when Uncle Francisco intervened and offered them the job. He gave them the same offer he made me: double what he paid the crew if they produced at least the same amount of work. The answer was always the same: no. They'd politely decline (with excuses) and leave. I now know why: it was a very difficult job. The relentless sun of the Andes mountains combined with dry weather carved indelible marks on the faces and bodies of the workers.

The crew Uncle Francisco employed was a great help to him and to others like him—otherwise, the farmers would see their crops rot due to the lack of employees. I would see this again when I moved to the United States many years later.

There are many aspects of life and difficult situations that we make harsh judgments against before knowing the facts.

Do not judge a person without having walked a mile in their shoes.

13. Discovering new business opportunities.

While working at the farm I learned about many of the vegetables that were grown there, noticing how different they were from the ones my parents bought from the neighborhood stores. I was impressed with the quality of Uncle Francisco's vegetables and not too long after this I had the inspiration to start my very first business: selling vegetables from the farm.

Keep your mind open to the possibilities around you. You never know where the inspiration for your "next great big idea" will come from.

14. Always get paid.

Don't give products or services away for free or people tend to think what you do or sell has no value. This is especially true for your services.

In the construction industry there is a concept that contractors should do estimates for free (as if their time, expertise, and other resources needed to produce the estimate would have cost nothing to acquire).. The industry has used free estimates as a way to get leads from work; however, the issue is the leads obtained by giving your services for free are most likely not good for you.

The sooner you learn who your customers are, the sooner you can come up with added value services that your customers will perceive as a good deal. Even if that service is just estimating, there is no reason why you need to do that for free. While many contractors would do that, you don't have to.

Find a way to offer something that will resonate with the people you want as customers and stick to it. You will get more qualified leads and close more deals than giving stuff away for free.

15. Price your products and services appropriately.

It is never based on how much you would pay for something. It is based on how much value your customers perceive the product or service has, which will determine how high or low you can charge.

16. Plan to succeed.

Don't only make contingent plans in the event something doesn't go the way you planned. Many people only make plans in the event things go wrong. Make plans to succeed so you can be ready to expand.

17. Compensate your people adequately.

This could be more than money, but ensure you do something that's fair and enticing to them. You need people to expand your business; otherwise you are limited in what you can do, and you'll never be able to expand or to have a proper enterprise.

18. Businesses are much more complex than childhood enterprises.

Pay close attention to details: regulations, laws, licensing, and taxes. I applaud you for wanting to start your own business, but never enlarge the statistics of failed businesses for lack of preparation.

Learn what you need before launching.

Prepare to succeed.

19. Never judge a book by its cover.

We never have an opportunity to learn from other people if we set them aside because of the way they look or what they do.

Exercise empathy with others, not judgment. Remember you can't really judge other people without having walked a mile in their shoes—and since you probably don't want to do that, reserve judgment.

20. Every other business provides good customer service.

Every customer also expects good service so the goal in your business should be to provide an *experience beyond your customers' expectations*. In Argentina return policies or warranties in the clothing business were unheard of. I challenged the idea but only because I could afford it.

Don't give your products or services away; price them according to the benefits your customer receives when they purchase them. Sell on value, never on price.

21. Be prepared for extreme situations.

New government regulations shut down my clothing business. The recent global pandemic shut down millions of businesses, thereby shattering lives.

You need to be prepared to pivot, find new opportunities, and to keep moving forward. There will always be challenges. Be prepared to conquer the challenges.

22. Focus.

You will achieve much more in life by focusing on one thing at a time. Pick one thing, and then plan, execute, correct, and repeat. No matter how small the task or goal you chose. Choose and get it done; then move on to the next.

23. Know why you need employees.

Before hiring your first employee, think carefully about what tasks this new person will help you accomplish and how that will produce more time for you to focus on other aspects of the business.

Never hire new employees because you assume you need them. You need to have specific tasks, and those tasks must be money-producing tasks, not busy-work-producing tasks.

24. Profits.

Put profits at the front and center of your business. Don't be ashamed of them. Profits and money are not dirty words.

Be proud of being a profitable business—your business can't function if you don't have profits.

25. Aim to be the best.

No matter what you want to be in life, aim to be the best of the best. Wherever you end up, aiming for second best is not an option.

26. Prepare for success, again.

I spoke about this before, but let me repeat it once again. Many businesses fail because they don't have plans if the best outcome happens—they are crushed by their own success. Don't make this mistake. Plans for undesired outcomes should always be secondary to plans for success.

27. Don't run away from your problems.

Problems will always find you. Face the music quickly so you can start on the road to recovery sooner. The faster you face your problems the faster you can move on.

28. Start.

Don't be afraid of taking the first step. Be afraid of the missed opportunities if you never did begin.

29. Regret and time.

I regret not taking more time to spend with friends and my extended family. Never make a choice that will cause regret. Make the time and do it now. Put your book down and call a friend, meet, go for coffee, lunch or a drink.

Why are you still reading? Put the book down and call a friend, *now*!

30. Sell on value not on price.

Selling on price rather than value ensures you will always be competing at the bottom. A business needs to be thriving rather than surviving.

31. Create differentiators.

Create differentiators between you and the competition. Customers have no other choice than to select based on price if you don't offer any differentiators. Every business is unique. Consider what you do differently than others and use that as a differentiator. It might be the services you provide or the *way* you provide them. Whatever it is, make sure your customers know why you are different and what you do differently than your competitors.

32. Learn your margin, markup, and overhead expenses.

Learn the associated costs of your business and find profit numbers that work for you, from there set your prices.

Do not base pricing on what others are charging. This doesn't mean you don't need to remain competitive, only that you need your own numbers. If your price ends up being higher, then you either need to become more efficient (lowering overhead), or come up with clear differentiators that justify the price difference.

33. Find a personal banker.

I cannot stress how important this is. I have been lucky to be able to establish a relationship with an excellent personal banker. Having an established relationship removes numerous financial stresses. Your personal banker is your personal champion inside the institution. He or she knows you and your business so they can use financial records to paint a holistic picture of your business rather than a cold spreadsheet. Personal bankers advocate for you and your businesses' best interests in front of the bank management and underwriters. This person will be fundamental in your success. Find this person soon.

34. Find an insurance agent.

It will greatly benefit you to find an insurance agent you can trust and who can offer sage advice. It is important to find an agent who has extensive experience in the construction industry and who is familiar with all the clauses and exclusions that may cause you trouble down the road.

Agents work on fees they get from the insurance companies, and the more policies you have with one agent the more you mean to his income, and therefore the more attention he can devote to you. Shop around for the best prices but always allow your agent to price a policy for you.

35. Know your numbers.

Understanding the financial numbers and statistics of your business is *critical* to your success!

If you don't understand these basic business formulas and numbers, and if you can't keep them under control, you will be out of business within a short amount of time

If you understand the numbers or if you are comfortable controlling them, you will be out of business in a short amount of time plus a day or two.

You must understand and control your numbers, and only then you have a chance to stay in business for many years!

Know your numbers!

36. Establish the right price.

Establishing the right price is critical to success, but don't mistake thinking higher prices are the solution.

For prices to be enticing to consumers they must provide lots of value. Your prices need to be competitive (but that doesn't mean lower than competitors). More importantly they need to provide value your customer sees and understands—if your customers cannot perceive the value they will only judge you based on price.

37. Explain price increases.

If your profit margins are declining, your first instinct will be to raise prices. This might be acceptable in market conditions like increased cost of materials. Be sure to explain to your customers why you are raising prices and then create added value for them.

For example, in the early 2020s materials costs increased by 20%, but at Smarter Remodeling we only adjusted our prices by 7%. Customers often don't realize your prices include much more than only materials, but they can see value by not being charged an additional 20%.

38. Understand company efficiency and overhead expenses.

If profit margins are declining, look for areas to improve efficiency in your company. Ask yourself, do I have too much overhead for the revenues I'm generating?

Lowering your overhead will allow you to increase your profit margins without raising prices.

39. Use technology to develop an online presence.

No matter how much you may or may not like technology and social media, the world has evolved and it's time to embrace these changes for your business.

If you don't want to run this aspect of your business, then consider outsourcing it to somebody you trust, but it is very important that you develop an online presence strategy. After all, like it or not, your potential customers are looking for you there.

40. Have a business, not a job.

Your aim should always be to have a business not just a job. It's fine to have a job inside your business but it's important not to be consumed by it.

Early on, when establishing your business work to create systems and procedures and hiring the right talent so your business can run without you.

This is not an easy process, but it's completely doable and has been done by millions of business owners who have already done it.

Remember, get started. It's one small step at the time.

41. Preplan and prepare for financing.

Regardless of your opinions of using banks or lenders to support your cash flow, it's better to have them lined up for an emergency situation than it is to be unprepared.

Set up your credit lines. Most credit options don't require you to use them, but do read the fine print, as occasionally some plans do require that they are used in order to keep them active.

42. Dictate customer payment terms.

This is your business. Why would you let a customer determine the way you want to run it? Never allow a customer to dictate the payment terms,much less allow a customer to get ahead of the game by performing more work than for how much money you have collected. This is your show, so get paid to perform.

43. Create your dream organization chart.

Spend time thinking about what positions your dream company will have. In the beginning, you probably will have your name in most boxes but over time this chart will help you hire the right people who fit into the right boxes. A business coach would be a great person to ask for assistance to complete this task.

44. Establish procedures and systems.

Create the procedures and systems so everybody in your company knows what to do at all times. I know it sounds like something only large companies do, but they are necessary if you intend on building a company with one or more employees.

Creating procedures and systems is a simple thing to do. Write down *how* you do things and *why,* so new employees can learn and do the same.

Start with something small. For example, write the procedure on how to open your office (alarms code, tricks with the lock (if any), where the lights are, etc.). Then write another. And then, another. Before you know it you will have many procedures on how to run your business.

45. Always be learning.

You don't have to be a student to learn. You can be a teacher and still learn lots of new things, from the material you prepared to the conversations that develop with your students or peers.

I've been surprised with the ideas that others have come up with when asked about certain situations.

There is always an opportunity to learn if you are open and willing. You can learn from your mistakes, you can get ideas from books or audiobooks, from watching videos, and from discussions with others.

There is always an opportunity to learn, so never stop looking for one. You never know where your inspiration for your next great idea will come from.

46. Learn to network.

At the beginning of my career I thought networking was an opportunity for me to sell to the people I met. I was mistaken. If you network with 20 people and your only goal is to sell them a product or service, then you will be done quickly as your pool of customers is very limited. But, if you invest your time in getting to know them, educating them on your business and the class of referrals you are looking for, you will create a salesforce of 20 people who can bring you more business than you ever dream of. Like a saying in BNI goes, "Networking, it's more about farming than hunting."

47. Think about profits.

Spend some time thinking about profits and the numbers your company needs to achieve its mission, but don't make the mistake of only raising prices to achieve that number. With each increase of price you need to provide an equal or greater amount of value so your customers can perceive they are still getting a deal.

Many contractors believe they can raise prices and customers will pay up. While this might be the case, they might also end up paying somebody else who could provide them with greater value.

Sometimes the solution is not in increasing prices, but rather making your company more efficient and reducing overhead costs.

48. Establish a pre-qualification process.

You need to establish a qualification process for every potential customer that comes your way. Determine the best method for you: a phone qualification, an in-person visit or a mix like the one Smarter Remodeling uses.

However, the following are critical:

1. Have a qualification process.
2. Charge for producing detailed proposals.

Don't give experience and time away for free.

49. Don't be hasty to disqualify potential customers.

Disqualifying customers on the phone without a pre-qualification process is a waste of advertising and marketing dollars.

Have a pre-qualification script to determine if this is a customer for whom you may want to do a quick site visit or assign to a salesperson.

At Smarter Remodeling, we charge for design consultations and detailed proposals but the ultimate goal is to qualify the potential customer, not to sell design services on a phone call. Have a qualification process and stick to it.

50. Know how to negotiate discounts.

Don't be so quick to offer a discount to close a deal. Many times customers just want something extra to feel they got the best part of the deal, but it doesn't mean you have to reach into your profits to give this. Be creative and be prepared.

51. Don't underestimate the help.

Never underestimate the power people lower on the totem pole hold. Treat everybody with respect and learn to listen.

Knowing who your customers are, what they want, and what they like can make the difference between walking away with a deal or with empty hands.

52. Make contacts.

There are no small contacts. Just contacts. And any of those can open the door to huge deals. Take good care of your contacts.

53. Keep your customers informed.

You have the power to change the outcome of the interaction regardless of how the other person approaches you. This is a very empowering thing to bear in mind.

Your behavior influences others' behavior, which is why it's crucial to employ certain skills to ensure a positive result.

It's your individual perception of "difficult" that defines the situation.

54. Don't label customers.

Don't allow employees to label a customer as "difficult." Encourage them to find what's making the customer appear as "difficult" and change the dynamics of the situation.

Allowing employees to label customers creates a pervasive culture of neglect.

55. Be proactive in providing great customer service.

You must be proactive to avoid uncomfortable situations with your customers. The vast majority of the time, customers don't want to have these conversations with you. But if you let small frustrations accumulate over time, you will have to deal with a problem of huge proportions, for which nobody knows the origin or how it got so big.

It can start with the first call you didn't make to update your customer on the progress or lack thereof. The smallest details matter, so make the call. Keep them informed even when there are no changes to report. It will save you many aggravations.

56. Go beyond customer service

Good customer service is largely expected by most customers. In order to differentiate from the competition, you need to go beyond this and delight your customers by creating excellent customer experiences. Your company culture should revolve around this and you should empower your entire team to create these experiences at every step of the process. From the initial call to project completion, everybody involved in the construction process should be trained and empowered to create unique, tailored experiences for each customer.

57. Use standardized guidelines to evaluate work.

Once more, don't be like Mr. Jones. Use standardized guidelines to define the quality of work your customer should expect. This will avoid confrontations with customers but, most importantly, you will one day win the coveted "Trim Installation of the Year" award. Wink, Wink.

58. Do not accept abusive behavior.

Do not engage in verbal matches with your customers. You cannot win by showing your wits and verbal skills. It's always better to let the customers calm down and then establish a conversation conducive to resolving the issue. If a conversation cannot happen due to the customer's unwillingness to talk, then why are you staying at that job? Terminate your contract using a clause like the one above and move on.

59. Be prepared for hard times.

Be prepared for difficult seasons, not only from a financial standpoint, but be ready to creatively pivot so you can survive an economic downturn—and even thrive.

We spend so much time lamenting over things we have no control over. Why don't we spend time coming up with ideas to change the things we can control? Like always, pick something small and get started. Don't just think about it anymore. Just do it.

60. Don't ever put work over family and friends.

A job is not worth losing family and friends. Your work can put you in a grave, but it won't shed a single tear for you. Your family and friends will. Family should always be first and friends a close second.

61. Lose the self-pity.

Don't drown yourself in self-pity because things are not going your way. Take action to change and to produce change. Things won't get better by themselves. You need to take an active role in making things happen for you.

62. Make everyone win.

The best deals are the ones where everyone wins. If you only make deals where only you win, you will be dealing with disgruntled parties long after the signing. You'll get much further ahead dealing fairly with others rather than taking advantage of them.

63. Price increases and contract clauses.

There will be times when it's necessary to increase prices due to unforeseen circumstances in a job and it's vital that there are clauses in your contracts to make allowances for this. Otherwise, you will not be able to do so. Check with your legal counsel to see if you can use something similar to the following clauses:

Material Price Escalation Clause: If, during the performance of this contract, the price of materials increases over 5%, though not the fault of the Company, the price shall be equitably adjusted by an amount reasonably necessary to cover any such significant price increases.

Reliance on Subcontractors Clause: Company relies upon subcontractors to perform the Work for the stated Contract Price. In the event that a subcontractor cannot perform any portion of the Work for the price quoted by the subcontractor, the Company, in its sole discretion, may cancel this Agreement or may increase the Contract Price if the cost to complete the Work quoted by the subcontractor exceeds 25% percent of the quote provided by the subcontractor and relied upon by the Company.

64. Pay yourself

Pay yourself.

I see many business owners making the mistake of working in their business and never collecting a paycheck. When asked why they respond they get paid with profits. If you are not paying yourself for the work you perform in your business then you might be masking an issue. When you might want to step aside your "profits" would now have to cover the salary of the person that would replace you. Are the profits enough to cover that salary and leave some left for the owner's compensation? If the answer is yes, then you would be fine. If the answer is no, you need to reevaluate your margins to increase your profits and consider start getting yourself paid.

65. Work with your spouse.

Take advantage of the opportunity you have to work with your spouse to spend some quality time together. Put your phones down at lunch unless it's to plan something together and try to engage in a conversation during the times you commute together. This would make for a much more enjoyable time at work.

66. Believe in yourself.

You have, inside you, all the strength you need to overcome any challenge. You just need to reach in, find it, and put it to work.

67. Train your employees.

Before hiring you should have the training material ready for the new employee. Hiring without this is just wasting hard earned money. Many contractors hire and assume the new hire will be trained by existing employees even when there is no formal training in place.

If your current employees are not properly trained, how do you expect the new ones to be?

Failing to make plans is planning to fail.

www.ingramcontent.com/pod-product-compliance
Lightning Source LLC
Chambersburg PA
CBHW070306200326
41518CB00010B/1907